11th Annual Larval Fish Conference

Major funding for the publication of these proceedings
was provided by the

Great Lakes Fishery Commission

and the

Michigan Department of Natural Resources

11th Annual Larval Fish Conference

Conference Sponsor

Department of Biological Sciences, Michigan Technological University

Nancy A. Auer, Conference Chair

Proceedings Sponsor

Early Life History Section, American Fisheries Society

11th Annual Larval Fish Conference

Edited by

Robert D. Hoyt

Proceedings of a Conference Held in
Houghton, Michigan, USA
June 1–3, 1987

American Fisheries Society Symposium 5

Bethesda, Maryland
1988

The American Fisheries Society Symposium series is a registered serial. Suggested citation formats follow.

Entire book

Hoyt, R. D., editor. 1988. 11th annual larval fish conference. American Fisheries Society Symposium 5.

Article within the book

Hildén, M., and L. Urho. 1988. Sampling of larval European smelt: a factorial experiment. American Fisheries Society Symposium 5:123–130.

Library of Congress Catalog Card Number: 88-72030

ISSN 0892-2284 ISBN 0-913235-57-1

Address orders to

American Fisheries Society
5410 Grosvenor Lane, Suite 110
Bethesda, Maryland 20814, USA

CONTENTS

Preface
Nancy A. Auer vii

Two Worlds for Fish Recruitment: Lakes and Oceans
John J. Magnuson 1

Description and Identification of Shortnose and Atlantic Sturgeon Larvae
Darrel E. Snyder 7

Spawning of Atlantic Herring in the Gulf of St. Lawrence
S. Messieh 31

Transplanting Pacific Herring Eggs in British Columbia: A Stocking Experiment
D. E. Hay and J. B. Marliave 49

Estimation of Hatch Periods for Yellow Perch, Based on Otolith Readings from Juveniles (Age-0)
Perce M. Powles and Stanley M. Warlen 60

Evidence for Baseline Flow Spikes as Spawning Cues for Colorado Squawfish in the Yampa River, Colorado
Thomas P. Nesler, Robert T. Muth, and Anthony F. Wasowicz 68

Variation in Pigment and Nape Morphology of Larval Tidepool Sculpin
Jeffrey B. Marliave 80

Marking Cyprinid Larvae with Tetracycline
Robert T. Muth, Thomas P. Nesler, and Anthony F. Wasowicz 89

Abundance, Distribution, Growth, and Mortality of Larval Lake Herring in Western Lake Superior
Jay T. Hatch and James C. Underhill 96

Movements of Young Lake Sturgeons Stocked in the Menominee River, Wisconsin
Thomas F. Thuemler 104

Spawning and Early Life History of Lake Sturgeon in the Lake Winnebago System, Wisconsin
James J. Kempinger 110

Sampling of Larval European Smelt: A Factorial Experiment
Mikael Hildén and Lauri Urho 123

Preface

The 11th Annual Larval Fish Conference was held at Michigan Technological University, Houghton, Michigan, June 1–3, 1987. It was sponsored by the Department of Biological Sciences. Fifty-nine delegates from five countries attended the conference. Thirty-three technical presentations were made during the three days of sessions. The keynote address, "Two worlds for Fish Recruitment: Lakes and Oceans," was given by John Magnuson, University of Wisconsin–Madison. The conference included two special sessions: "North American Sturgeon Larvae" was directed by Fred P. Binkowski, and "Lost and Not Found Populations" was chaired by Joseph Graham. Other general session chairs were Pamela Mansfield, John Aho, and Robert Muth. An afternoon workshop included displays of educational materials for educators, by Darrel Snyder; of sampling gear, by Oceanics; and of computer imagery, by BioSonics.

The Sally Richardson Best Paper Award was presented to Mikael Hildén and Lauri Urho for their paper "Sampling of Larval European Smelt: A Factorial Experiment." Conference activities included a picnic lunch on the shore of Lake Superior, a Finnish banquet, and an impromptu field trip to the Sturgeon River to seine for lake sturgeons and to examine spawning habitat.

Major funding for publication of papers in this issue was provided by the Great Lakes Fishery Commission, Ann Arbor, Michigan, and the Michigan Department of Natural Resources, Lansing.

NANCY A. AUER, *Conference Chair*
Department of Biological Sciences
Michigan Technological University
Houghton, Michigan 49931

Editorial Acknowledgments

The Editorial Board for the 11th Conference thanks the following manuscript reviewers for their contributions to these proceedings.

Nancy A. Auer
Robert Behnke
Fred Binkowski
Chris Brown
Larry J. Buckley
Villy Christensen
Larry Crowder
Charles C. Coutant
Michael P. Fahay
Kenneth Frank
Robert Fridley
J. R. Goulet, Jr.
Joseph Graham
J. R. Green
M. D. Groslein
T. Derrick Iles
Kim Jauncy
Lynn Kaeding
James Kitchell
David Kohlhorst
Ken C. Lindeman
R. Gregory Lough
Joanne Lyzckowski–Shultz
John Magnuson
Bruce Manny
Robert W. McCauley
William J. McNeil
Chris A. Mills
Robert J. Muncy
Robert Muth
Steven Murawski
John G. Nickum
Dietrich Schnack
Thomas P. Simon
Darrel E. Snyder
David Stevenson
David Townsend
Richard Valdez

American Fisheries Society Symposium 5:1–6, 1988

Two Worlds for Fish Recruitment: Lakes and Oceans

JOHN J. MAGNUSON

*Center for Limnology and Department of Zoology, University of Wisconsin-Madison
Madison, Wisconsin 53706, USA*

Abstract.—Lakes and oceans have many similarities for the ecology of larval and adult fishes, but they differ in at least one important respect. Lakes have characteristics of islands, poorly connected with one another and subject to locally extreme environmental conditions. These circumstances inhibit immigration of fishes and increase their chances of extinction. Adjacent ocean patches, in contrast, are broadly connected and environmentally similar, conducive to immigration but not to extinction. At similar spatial scales, then, recruitment of species and year classes may be influenced principally by extinction processes in lakes and by colonization processes in oceans.

Interests in larval fish ecology span two of the world's major waterscapes—lakes and oceans. Most scientists who study recruitment work in lakes *or* in oceans, but not in both. Does this distinction simply indicate a pragmatic difference in the choice of field sites, or are there fundamental differences in the ecology of fish recruitment in lakes and oceans?

Just how similar or how different are lakes and oceans in respect to characteristics fundamental to larval fish biology and recruitment? When the Society of Limnology evolved to become the Society of Limnology and Oceanography, Alfred C. Redfield, an oceanographer interested in physiology and chemistry, wrote in the announcement of the journal *Limnology and Oceanography* (1956, volume 1, number 1) that "Just as it had been accepted from the start that students of the biological, chemical and physical aspects of limnology had much to gain by association, so it was soon realized that the differences between fresh and salt water systems were trivial, when compared to the common principles with which limnologists and oceanographers alike are concerned."

Some of us, when we compare things, see everything as different, and we stress dissimilarity. Others of us tend to see everything as the same, and we stress similarity. Certainly there is an infinite number of differences as well as an infinite number of similarities between lakes and oceans. To me, the similarities are the fundamental process-oriented things. For example, oxygen is relatively rare, and important sets of adaptations are apparent during the egg and larval stages for providing sufficient oxygen. Pelagic eggs and larvae are advected by currents. Buoyancy and vertical mixing are important both to the larvae and to their food supply. Fishes are ectotherms, and temperature and seasonality are major environmental factors. Fish are size-selective feeders. Yolk-sac absorption is often a critical process. Predation and cannibalism are important mortality factors. Mortality rates are high.

The differences seem to be the small, fascinating details of specific cases. The density of water is a little different: 1.000 g/mL for fresh water versus 1.025 g/mL for seawater. Fresh water can hold a little more dissolved oxygen than seawater. Different species of fish and their prey abound in the two environments. Many of the differences we could call to our attention are no more than those found between different lakes or different parts of the sea. Some of the assumed differences do not hold up in any absolute sense. For example, lakes are fresh and oceans are salty but many inland waters are brackish or even hypersaline, like Great Salt Lake in Utah, and many parts of the ocean are brackish or almost fresh, like Bothnian Bay in the Baltic Sea.

The similarities provide the matrix in which larval ecology operates; the differences provide the interesting specific cases scientists study. Members of the Early Life History Section of the American Fisheries Society are drawn together by the similarities of larval fish biology and ecology among all aquatic environments. As individuals study specific problems of specific taxa in specific environments, they bring the interesting variability to each other to test against the underlying questions of larval fish biology and recruitment. Ocean versus lake would not appear to be a fundamental dichotomy for their science and understanding.

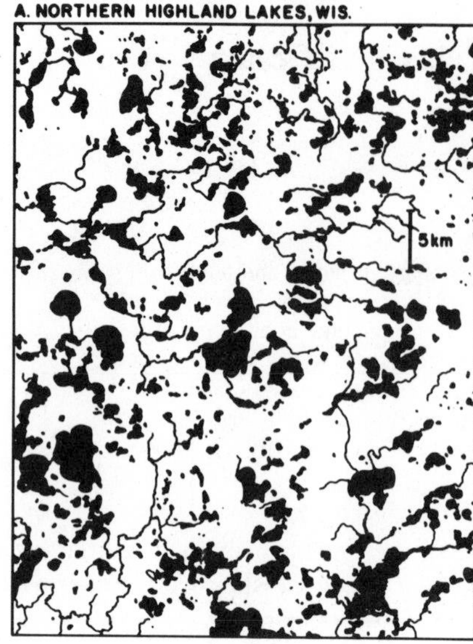

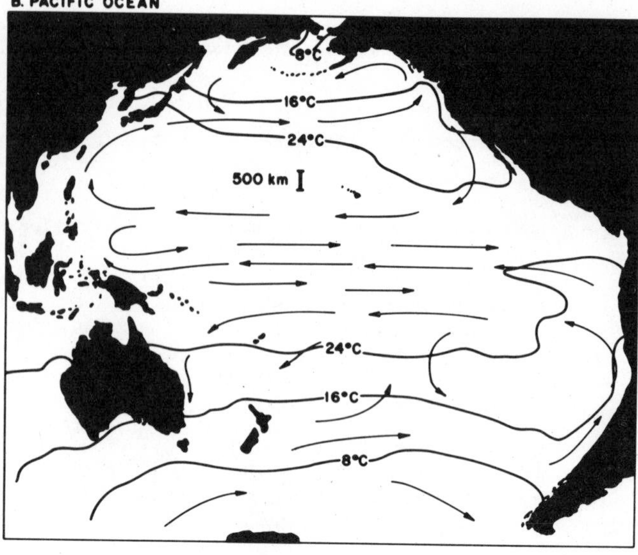

ISLAND ARCHIPELAGO

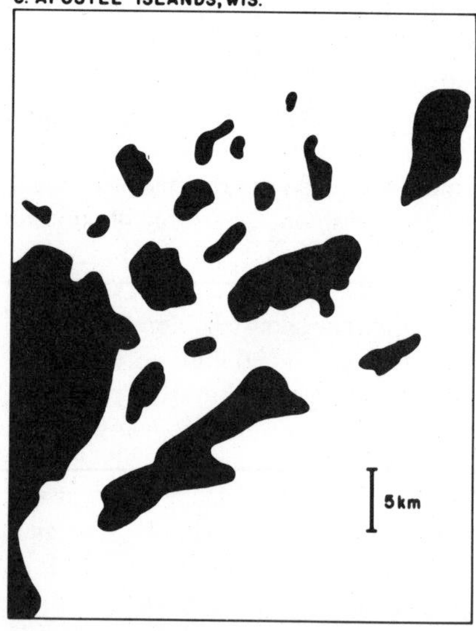

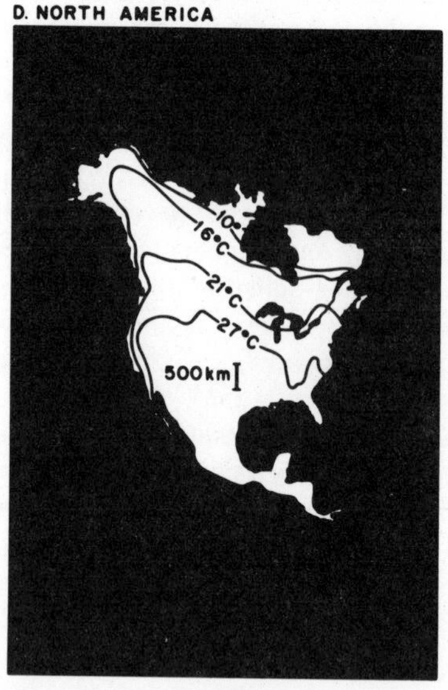

FIGURE 1.—Maps of a lake district, an ocean, an island archipelago, and a continent to emphasize the small size and isolation of lakes and islands and the large size and connectedness of oceans and continents. The Northern Highland lake district is in northern Wisconsin. The Pacific Ocean map is modified from Bramwell (1977) and shows July mean surface temperatures. The Apostle Islands are along the Wisconsin shore of Lake Superior. The North America map is modified from Espenshade (1984) and shows July normal temperatures. Scales are approximate.

Is there, then, no fundamental difference between lakes and oceans that provide a different matrix for larval ecology? I believe that there is, and, although I do not understand the implications very well, the difference is very simple. Thoughts on this fundamental difference between lakes and oceans are flagged by an analogy with terrestrial ecology—lakes are like islands; oceans are like continents (Figure 1).

Lakes, like islands, are relatively small, isolated, geologically young, and rather temporary ecosystems. Oceans, like continents, are rela-

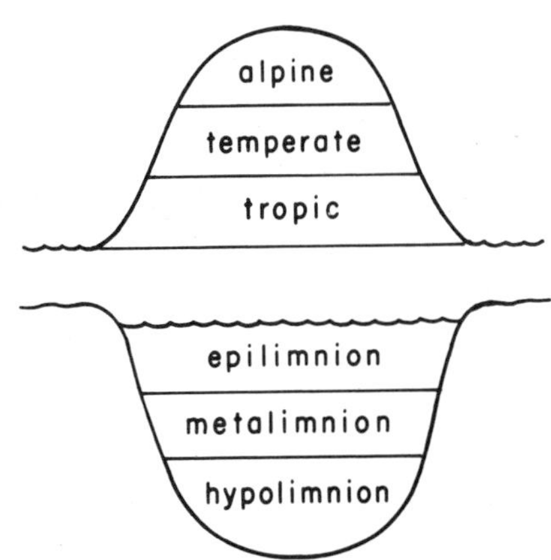

FIGURE 2.—Vertical thermal zonation of lakes and oceans versus that of islands and continents. Lakes and oceans are like islands and continents turned upside down. The epilimnion or uppermixed layer is warm in summer (or in the tropics) and is analogous to the warm, low-altitude regions of islands and continents in summer (or in the tropics). The metalimnion or thermocline is cool and analogous to intermediate, more temperate altitudes of islands or continents in summer (or in the tropics). The hypolimnion is deep, cold water analogous to the high-altitude, alpine regions of islands and continents in summer (or in the tropics).

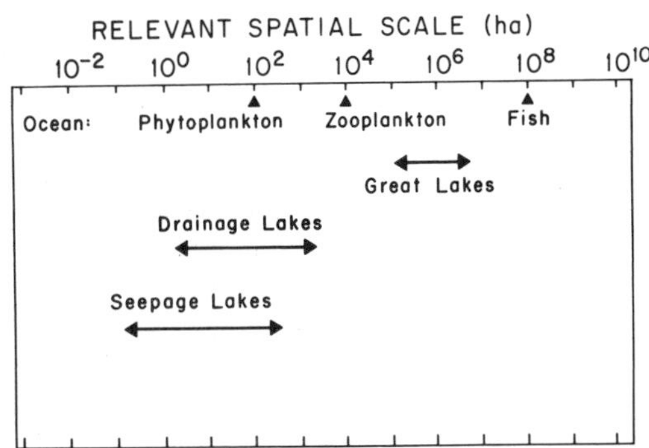

FIGURE 3.—Comparison of spatial scales for large-scale patchiness of biota in the ocean (Steele 1977) with the surface areas of great lakes, smaller drainage lakes with stream outlets, and even smaller seepage lakes without outlets in the Northern Highland lake district of Wisconsin. The great lakes include those as large in area as Lake Superior and as small as Lake Erie. One hectare (ha) equals 10,000 m^2.

tively large, connected, geologically old, and rather permanent. Lakes are imbedded in a matrix of land, as islands are in a matrix of water. Oceans are their own matrix, as continents are their own. An ocean is an open environment dominated by currents and mixing processes with relatively fuzzy boundaries such as thermal gradients among its parts. A lake district is constituted as a set of relatively closed basins often separated by sharp land boundaries or more fuzzy lotic boundaries. If anything, continents are intermediate between lakes and oceans in respect to their openness.

Lakes and oceans are a bit like islands or continents turned upside down (Figure 2). Islands and continents are warm near their base and cold at high altitudes. Lakes and oceans are warm near the surface and cold at depth. Many other analogies can be gathered; see Barbour and Brown (1974), Magnuson (1976), and Eadie et al. (1986) for additional analogies between lakes and islands.

A fundamental difference between lakes and oceans may result simply from the large size of the ocean versus the small size of a single lake (Figure 1). Steele (1977, 1984) has suggested scales relevant to major patchiness for fishes, zooplankton, and phytoplankton in the sea. When these spatial scales are compared with absolute sizes of lakes (Figure 3), it can readily be observed that the posited scale for ocean fishes does not even fit into the largest lakes of the world, Lake Superior and Lake Baikal. Many northern Wisconsin lakes are five to nine orders of magnitude smaller than the scale for ocean fishes and even three orders of magnitude smaller than the posited scale for phytoplankton patchiness in the sea.

The fit is not quite as impossible for larval fishes and fish eggs as it is for adult fish because, at least for the short term, larval fishes would be expected to operate on scales as small as do zooplankton. Clearly, many fishes flourish in these small-scale inland waters. There are about 8,400 fish species normally living in freshwater lakes and streams and about 13,200 living entirely in the oceans, even though only 0.01% of the world's surface waters are fresh water (Nelson 1984). Yet, one can not help but wonder whether certain fundamental scale-dependent processes important in the sea just do not fit into small-scale lake ecosystems. Certainly the very largest structures in physical oceanography such as the Gulf Stream or the North Pacific Gyre do not fit; if fishes have adaptations to the large-scale features of such phenomena, these adaptations would not be needed in inland waters.

The spatial scales most important to fish larvae are probably the ones of millimeters to meters, as is apparent in Kerfoot and Sih (1987). These are the scales at which prey capture and predator avoidance occur. At the north edge of the Gulf Stream, patches of 70-kHz acoustic scatterers, which include ichthyoplankton, macrozooplankton, and micronekton, can be as small as 2 m vertically though 50 m horizontally (R. W. Nero, University of Wisconsin-Madison, unpublished data). Small-scale vertical patchiness of prey for pelagic larvae determines the success or failure of a year class for some pelagic fishes (Lasker 1975; Peterman and Bradford 1987). Apparently, the spatial heterogeneity important to processes determining life and death of fish eggs and larvae occurs at small scales in both lakes and oceans. I question whether the size difference per se between lakes and oceans leads to fundamental differences in the ecology of fish recruitment.

I also doubt that the age and permanence of the ocean and the youth and temporary nature of lakes has led to any fundamental differences between the ecology of fish larvae in lakes and oceans. Although individual lakes in their present arrangement are temporary, lakes as a class of environments are continuous in history. They are destroyed and re-created from time to time by climatic processes such as glaciation and drought and by geologic processes such as tectonic movements and erosion or deposition. For example, there is no trace of the small lake basins that probably existed in the northern highlands of Wisconsin and Michigan prior to the last Wisconsin glaciers, which receded about 10,000 years ago, but the fish species that lived in the region have been displaced southward during each glacial advance and have reinvaded each time the continental glaciers receded. The fish genome and the larval adaptations it carries have been more persistent in time than have the individual lakes the fish species inhabit.

If there is a fundamental difference in fish ecology between lakes and oceans owing to their respective island and continental features, I suggest that it originates from the closed, isolated nature of lakes versus the open, connectedness of an ocean. Firstly, the species composition of a lake in ecological time is the net result of invasions and extinctions of species (Magnuson 1976). The same should be true for small pieces of the ocean with the added complications that free-ranging transients may immigrate and emigrate, and that the piece of ocean may be defined either as a geographic space or a moving patch of water. But invasion events are infrequent in lakes; it may take thousands of years for all the species of a region to gain access to all parts of a lake district. In contrast, all species in a region of the ocean probably access the entire piece of ocean within months or certainly within a few years. Secondly, adjacent lakes tend to be more environmentally heterogeneous than adjacent small pieces of ocean, which tend to be more homogeneous owing to physical mixing processes. An extinction factor such as low pH or low oxygen may be present in only one of two adjacent lakes (for examples, see Tonn and Magnuson 1982). Thirdly, such extinction factors often take their toll within a few years after a species invades (for examples, see Magnuson et al., in press). Owing to these three phenomena, the long-term spatial distribution of fish species among lakes in a lake district should have the mark of extinction processes, whereas the distribution in comparably sized pieces of the ocean should have the mark of the colonization events.

In the lake district pictured in Figure 1, the species composition of small forest lakes is best predicted by extinction characteristics of the lakes rather than by differences in the accessibility of the lakes to fishes (Tonn and Magnuson 1982; Rahel 1984). This is so even though many of the lakes have no water connection to adjacent lakes. Extinction factors are intense in these lakes: low pH in deep isolated bog lakes, low dissolved oxygen in winter in shallow lakes, and the presence of large piscivorous fishes in lakes physically benign enough to support them. All three of these cause local extinction within a few years after an invasion or stocking.

I know of few data on the colonization and extinction of fishes in pieces of ocean the size of small forest lakes. The role of recruitment from up-current areas was documented by Cowen (1985) for the coastal area of California. Sale (1980) and Sale and Douglas (1984) pointed out that recruitment events, that is, invasion of ichythoplanktonic larvae, impart continuous change to the assemblages of fishes on coral patch reefs at small spatial scales. In these cases, the presence of a species is very much determined by variation in recruitment related to the open nature of the ocean. I have not seen comparable data for open water, but pelagic fish larvae are patchily distributed, and local extinctions owing to predation, starvation, and advection are thought to contribute to these distributions (Sharp 1981; Rothschild and Rooth 1982; Sharp and Csirke 1983; Doherty et al. 1985). I suspect that such local extinctions in the open ocean are soon erased by invasion of new recruits from other locations.

Certainly there are parts of the ocean edge and frontal zones that have some degree of isolation and there are lake districts that are more open and connected than others. The relative importances of invasion and extinction events in determining species composition should change along these gradients from more closed to more open environments.

I have enjoyed speculating about fundamental differences between lakes and oceans that may establish distinctly different matrices for larval ecology and the recruitment of fishes. I think that comparisons between lakes and oceans and the study of the same processes at the same spatial scales in lakes and oceans can lead to new insights. Although the processes that cause recruitment to vary appear to be similar in both lakes and oceans and occur over small spatial scales in both waterscapes, I suggest that the island nature of lakes and the continental nature of oceans provide fundamentally different contexts for fish recruitment. This difference in connectedness between groups of lakes and adjacent small pieces of ocean leads to different probabilities of local extinction and different probabilities of local recolonization. Thus, for the same spatial scale, recruitment in lakes may be best predicted by extinction models whereas recruitment in oceans may be best predicted by colonization models.

Acknowledgments

I thank Nancy A. Auer, Fred P. Binkowski, and the Early Life History Section of the American Fisheries Society for inviting me to prepare this paper. Among the many colleagues and students who have influenced my thoughts on this topic are Timothy F. H. Allen, Stephen B. Brandt, Peter F. Sale, and William M. Tonn. I thank the National Science Foundation, the Office of Naval Research, Wisconsin Sea Grant, the Environmental Protection Agency, the former Bureau of Commercial Fisheries, and the University of Wisconsin for providing the opportunities for me to observe lakes and oceans as well as islands and continents.

References

Barbour, C. D., and J. H. Brown. 1974. Fish species diversity in lakes. American Naturalist 108:473–489.

Bramwell, M., editor. 1977. The Rand McNally atlas of the oceans. Rand McNally, New York.

Cowen, R. C. 1985. Large scale pattern of recruitment by the labrid, *Semicossyphus pulchur*: causes and implications. Journal of Marine Research 43:719–742.

Doherty, P. J., D. M. Williams, and P. F. Sale. 1985. The adaptive significance of larval dispersal in coral reef fishes. Environmental Biology of Fishes 12:81–90.

Eadie, J. M., T. A. Hurly, R. D. Montgomerie, and K. L. Teather. 1986. Lakes and rivers as islands: species–area relationships in the fish faunas of Ontario. Environmental Biology of Fishes 15:81–89.

Espenshade, E. B., Jr., editor. 1984. Goode's world atlas, 16th edition. Rand McNally, Chicago.

Kerfoot, W. C., and A. Sih, editors. 1987. Predation. Direct and indirect impacts on aquatic communities. University Press of New England, Hanover, New Hampshire.

Lasker, R. 1975. Field criteria for survival of anchovy larvae: the relation between inshore chlorophyll maximum layers and successful first feeding. U.S. National Marine Fisheries Service Fishery Bulletin 73:453–462.

Magnuson, J. J. 1976. Managing with exotics—a game of chance. Transactions of the American Fisheries Society 105:1–9.

Magnuson, J. J., C. A. Paszkowski, F. J. Rahel, and W. M. Tonn. In press. Fish ecology in severe environments of small isolated lakes in northern Wisconsin. *In* R. Sharitz and J. Gibbons, editors. Freshwater wetlands and wildlife. University of Georgia, Savannah River Ecology Laboratory 4th Symposium, Aiken, South Carolina.

Nelson, J. S. 1984. Fishes of the world. Wiley, New York.

Peterman, R. M., and M. J. Bradford. 1987. Wind speed and mortality rate of marine fish, the northern anchovy (*Engraulis mordax*). Science (Washington, D.C.) 235:354–356.

Rahel, F. J. 1984. Factors structuring fish assemblages along a bog lake successional gradient. Ecology 65:1276–1289.

Rothschild, B. J., and G. H. Rooth, conveners. 1982. Fish ecology III. A foundation for REX a recruitment experiment. University of Miami, Technical Report 82008, Miami.

Sale, P. F. 1980. The ecology of fishes on coral reefs. Oceanography and Marine Biology: an Annual Review 18:367–421.

Sale, P. F., and W. A. Douglas. 1984. Temporal variability in the community structure of fish on coral patch reefs, and the relation of community structure to reef structure. Ecology 65:409–422.

Sharp, G. D., editor. 1981. Workshop on the effects of environmental variation on the survival of larval pelagic fishes. Intergovernmental Oceanographic Commission, Workshop Report 28, Paris.

Sharp, G. D., and J. Csirke, editors. 1983. Proceedings of the expert consultation to examine changes in abundance and species composition of neritic fish resources. FAO (Food and Agriculture Organization of the United Nations) Fisheries Reports 291, volumes 1–3.

Steele, J. H. 1977. Some comments on plankton patches. Pages 1–20 *in* J. H. Steele, editor. Spatial pattern in plankton communities. NATO (North Atlantic Treaty Organization) Conference Series 4: Marine Sciences, volume 3.

Steele, J. H. 1984. Kinds of variability and uncertainty affecting fisheries. Pages 245–262 *in* R. May, editor. Exploitation of marine communities. Springer-Verlag, New York.

Tonn, W. M., and J. J. Magnuson. 1982. Patterns in the species assemblages in northern Wisconsin lakes. Ecology 63:1149–1166.

American Fisheries Society Symposium 5:7–30, 1988

Description and Identification of Shortnose and Atlantic Sturgeon Larvae[1]

DARREL E. SNYDER

Larval Fish Laboratory, Colorado State University, Fort Collins, Colorado 80523, USA

Abstract.—Larvae of shortnose sturgeon *Acipenser brevirostrum* and Atlantic sturgeon *Acipenser oxyrhynchus*, cultured and collected from various sources, were examined for morphometric, meristic, pigmentary, and size-related characters to document morphological development and facilitate diagnosis. Shortnose sturgeons typically hatch at 9–10 mm standard length (SL), complete yolk absorption by 14 mm SL, acquire their first fin rays and scutes between 21 and 24 mm SL, and acquire the full complement of fin rays, except in the caudal fin, by 57 mm SL. Atlantic sturgeons typically hatch at 7–9 mm SL, complete yolk absorption by 13 to 14 mm SL, acquire their first scutes between 17 and 20 mm SL, acquire their first fin rays at 21 mm SL, and acquire a full complement of fin rays, except in the caudal fin, between 47 and 58 mm SL. Mean myomere counts for both species are 38 preanal and 22 or 23 postanal. Most recently hatched larvae can be tentatively identified by size relative to state of development and, depending on the river, by collection date and location. In some rivers, shortnose sturgeons spawn about a month earlier and farther upstream than Atlantic sturgeons. Following yolk depletion, ventral pigmentation and distance between lobes of the lower lips are the most obvious diagnostic characters. The ventrolateral and ventral surfaces of the abdomen are white on shortnose sturgeons but covered with melanophores on Atlantic sturgeons, except on the midventral surface of smaller specimens. The distance between the two lobes of the lower lip is greater than 25% of mouth width (including lips) for shortnose sturgeons and less than 20% of mouth width for Atlantic sturgeons. For specimens over 60 mm SL, shortnose sturgeons also have 17–22 pelvic and 18–24 anal fin rays, whereas Atlantic sturgeons have 26–33 pelvic and 22–30 anal fin rays.

The Atlantic coast of North America from the Saint Johns River in Florida to the Saint John River in New Brunswick is inhabited by both the shortnose sturgeon *Acipenser brevirostrum* and the Atlantic sturgeon *Acipenser oxyrhynchus oxyrhynchus* (Lee et al. 1980; Dadswell et al. 1984; Smith 1985). Atlantic sturgeon also range north into and above the Gulf of Saint Lawrence, and, as another subspecies *Acipenser o. desotoi*, from the Mississippi Delta to the west coast of Florida and elsewhere in the Gulf of Mexico.

The shortnose sturgeon is protected as an endangered species in the USA, yet the larger and more commercially valuable Atlantic sturgeon has become nearly as rare in many localities. In some South Carolina waters, shortnose sturgeons may actually outnumber Atlantic sturgeons (T.I.J. Smith, Marine Resources Research Institute, personal communication). Atlantic sturgeons are given legal protection only in the states of Connecticut, Rhode Island, New Jersey, Florida, and Mississippi (Johnson 1987).

To better understand these fish, prevent further detrimental impact on their populations due to habitat modifications and pollution, and attempt recovery efforts, it is critical to identify spawning seasons, spawning and nursery grounds, and habitat or ecological requirements of spawning adults, eggs, and larvae. Collections of fish larvae are necessary for gathering much of this information.

As with many other species of North American fishes, the larvae of these sturgeons have been inadequately described for taxonomic purposes. Ryder (1890) described the development of Atlantic sturgeon larvae based on specimens reared from the Delaware River (Delaware) stock but, with one exception, illustrated his description with previously published drawings of European species (*A. ruthenus* and *A. huso*). The exception, a drawing reproduced in larval fish manuals by Mansueti and Hardy (1967), Lippson and Moran (1974), and Jones et al. (1978), is reported to be a just-hatched Atlantic sturgeon measuring 11.5 mm total length (TL). In apparent contradiction, Smith et al. (1980), who documented the early behavior and growth of hatchery-reared Atlantic sturgeon from South Carolina, reported a mean hatching size of 7.1 mm TL.

Most of the descriptive information published

[1]Contribution 34 of the Larval Fish Laboratory, Colorado State University.

on shortnose sturgeon larvae, as well as other aspects of shortnose sturgeon biology, was summarized by Dadswell et al. (1984). Pekovitch (1979) provided a few descriptive comments, a table of selected measurements and myomere counts for four shortnose sturgeon larvae between 16.3 and 18.2 mm TL, and three-view drawings of specimens from the Hudson River (New York) that were 16.3 and 32.0 mm TL. Taubert and Dadswell (1980) described shortnose sturgeon larvae collected from the Holyoke Pool of the Connecticut River (Connecticut) and the Saint John River and included photographs of specimens that were 10.0 and 14.7 mm TL. Buckley and Kynard (1981) reported on the behavior of shortnose sturgeon larvae reared from Holyoke Pool stock and published photographs of 2-h-, 1-d-, 4-d-, and 8-week-old specimens.

Dovel (1979) noted that shortnose sturgeons as small as 17 mm TL can be distinguished from Atlantic sturgeons by features of the mouth. His report included ventral-view photographs of the heads of 17-mm specimens of both species but failed to specify the differences. Taubert and Dadswell (1980) suggested that, after the mouth is formed, mouth width (diagnostic for adults) might be useful for separating shortnose from Atlantic sturgeon larvae. Bath et al. (1981) documented the value of mouth width as a diagnostic character and described larvae of both species from the Hudson River (New York). They provided morphometric and meristic data for 26 larvae (8.4–37.0 mm TL) and photographs of seven specimens (8.4–31.5 mm TL). Ratio of mouth width to head width, degree of development (not explained), collection site, and date of capture were found useful for identifying most non–yolk-bearing larvae. Yolk-bearing larvae were tentatively identified according to location and date of capture. Dovel and Berggren (1983) reprinted the ventral head photographs from Dovel (1979) and noted that "while there are differences in pigmentation, shape and size of barbels and features of the mouth, the relative proportion of the space between the fleshy lobes of the mouth to the width of the mouth appears to be a valid distinguishing characteristic. . . ."

The purposes of this paper are to (1) more completely document the morphological development of shortnose and Atlantic sturgeon larvae, (2) confirm and elaborate on the value of previously noted criteria for diagnosis, and (3) reveal additional characters that may be of taxonomic value.

Methods

Most specimens examined for this study were reared or collected in South Carolina (Table 1). However, to expand upon the available size range of naturally spawned specimens and assure that diagnostic criteria derived from this study are applicable to other populations of Atlantic and shortnose sturgeons, additional specimens were solicited for loan from other east coast researchers; most specimens received came from the Hudson and Saint John rivers.

All cultured specimens from the Orangeburg National Fish Hatchery, except 29–39-mm SL Atlantic sturgeons, were preserved in formalin solutions. The remaining specimens were variously preserved in formalin or alcohol solutions (Table 1). Resultant measurements reported here may reflect varying and sometimes considerable degrees of shrinkage, possibly as much as 10% for some very early larvae in alcohol solutions (e.g., the smallest probable Atlantic sturgeon larva from the Savannah River). Shrinkage is much less for specimens preserved in formalin than in alcohol.

Based on definitive diagnostic characters derived from cultured specimens during this study, collected specimens over 12 mm SL were identified or verified as either shortnose or Atlantic sturgeon. Smaller yolk-bearing larvae were tentatively identified or verified on less-certain criteria derived from cultured specimens, as well as information on collection sites and dates. Data from collected material were used to complement and substantiate observations based on cultured material.

Specimens were examined for differences in 52 measurements, 18 counts, external morphology, melanophore pigmentation, and developmental state relative to size under a stereozoom microscope (3.5–30× magnification) with various combinations of reflected and transmitted light. An eyepiece reticle was used for direct measurement of specimens to the nearest 0.05 or 0.1 mm. Repeatability of measurements was within 0.1 mm. Questionable measurements (i.e., those that appeared inconsistent with the same measurements on other, similar-sized specimens) were rechecked for verification or correction. When feasible and not reported by the authors, similar characters were also observed, measured, or counted from published illustrations for comparative purposes.

The specific measurements used in this investigation are diagrammed in Figure 1 and expand upon those detailed by Snyder (1981, 1983). Un-

TABLE 1.—Specimens of shortnose and Atlantic sturgeon larvae examined.[a]

Number examined	Standard length (mm)	Source	History	Preservative	Depository
			Shortnose sturgeon		
24[b]	9–24	ONFH	Cultured; parent stock from Cooper R., SC; hatched Mar 9, 1983; preserved Mar 9–Apr 3, 1983	Formalin	Returned to ONFH
3	15–16	SABS	Cultured; parent stock from Saint John R. near Fredericton, NB; hatched May 25, 1980; preserved Jun 24, 1980	Alcohol	SABS, 678-7
3	9–10	SRL–ECS	Collected from Savannah R., SC, between km 113 and 253, Mar 12 and 26, 1982 and Mar 22, 1980	Alcohol	Returned to ECS
6	9–11	SRL–ECS	Collected from Savannah R., SC, between km 113 and 253, Mar 22–29, 1983	Formalin	Returned to ECS
1	15	LES	Collected from Hudson R., near Albany, NY, km 235, May 21, 1979	Formalin	LFL
10	15–17; 34–57	NYSDEC–HES	Collected from Hudson R., NY, km 187–245, 1977–1980	Alcohol	AMNH
2	<50–51	SCWMRD	Removed from stomach of a yellow perch *Perca flavescens* caught in Merrymeeting Bay near New Brunswick, Maine, Jun 30, 1975	Formalin	Returned to SCWMRD
			Atlantic sturgeon		
20[b]	9–17	ONFH	Cultured; parent stock from Edisto R. and Combahee R., SC; hatched Mar 30, 1981; preserved Apr 12, 1981	Formalin	Returned to ONFH
8[b]	29–39	ONFH	As above but preserved Apr 21, 1981; five specimens cleared and stained for cartilage	Alcohol	Returned to ONFH, except five[b] retained by LFL
9	58–136	ONFH	Cultured; parent stock from Atlantic Ocean off north Georgetown jetty, SC; hatched Mar 1, 1979; preserved Jun 11–14 and Jul 21, 1979	Formalin	Returned to ONFH
3	6–8	SRL–ECS	Collected from Savannah R., SC, between km 113 and 253, May 21 and Aug 12, 1982	Alcohol	Returned to ECS
6	8–14; 47	SRL–ECS	Collected from Savannah R., SC, between km 113 and 253, Apr 26–May 18 and Jun 14, 1983	Formalin	Returned to ECS
1	9	SABS	Collected from Saint John R. near Fredericton, NB, Jun 2, 1980	Alcohol	SABS, 864-7
4	14–31	LES	Collected from Hudson R., NY, km 64 to 90, Jul 2, 1968, Jun 22 and 29, 1976; 20-mm specimen bleached or cleared	Formalin	LFL
1	47	SCWMRD	Collected from Winyah Bay near Georgetown, SC, Jun 1978 (collection 78108, project 01)	Alcohol	Returned to SCWMRD

[a]Abbreviations: AMNH = American Museum of Natural History, NY, NY; ECS = Environmental and Chemical Sciences, Incorporated, Aiken, SC; HES = Hazelton Environmental Sciences Corporation, Northbrook, Illinois; LES = Laboratory for Environmental Studies, New York Medical Center, Tuxedo, NY; LFL = Larval Fish Laboratory, Colorado State University, Fort Collins, Colorado; NB = New Brunswick; NY = New York; NYSDEC = New York State Department of Environmental Conservation, Stamford; ONFH = Orangeburg National Fish Hatchery, SC; R = River; SABS = Saint Andrews Biological Station, NB; SC = South Carolina; SCWMRD = South Carolina Wildlife and Marine Resources Department, Charleston, SC; SRL = Savannah River Laboratory, Aiken, SC.

[b]Additional specimens were available but not specifically measured or analyzed.

less specified as total length (TL), specimen length is reported as standard length (SL) and measured to the end of the notochord. The notochord persists even in adults nearly to the end of the heterocercal tail. Most proportional measurements are reported herein to the nearest whole unit as a percentage of standard length (%SL). Percent total length (%TL) was not used because the membranous end of the caudal fin of some specimens was damaged and required estimation of TL. However, the difference between SL and TL for sturgeons is usually very small, so %SL closely approximates %TL.

Counts made in this investigation also followed those detailed by Snyder (1981, 1983) with some modifications and additions. Only preanal (to posterior margin of vent), postanal, and total myomeres were counted (Figure 1). Preanal myomere counts included the anterior segment immediately behind the region of the occiput (stippled in Figure 1) and all myomeres even partially transected by a vertical line from the posterior margin of the vent (Seifert 1969; Snyder 1981; Fuiman 1982). The first few myomeres are epaxial, and the most anterior vertical myoseptum often appears partial or incomplete. Fin ray counts in-

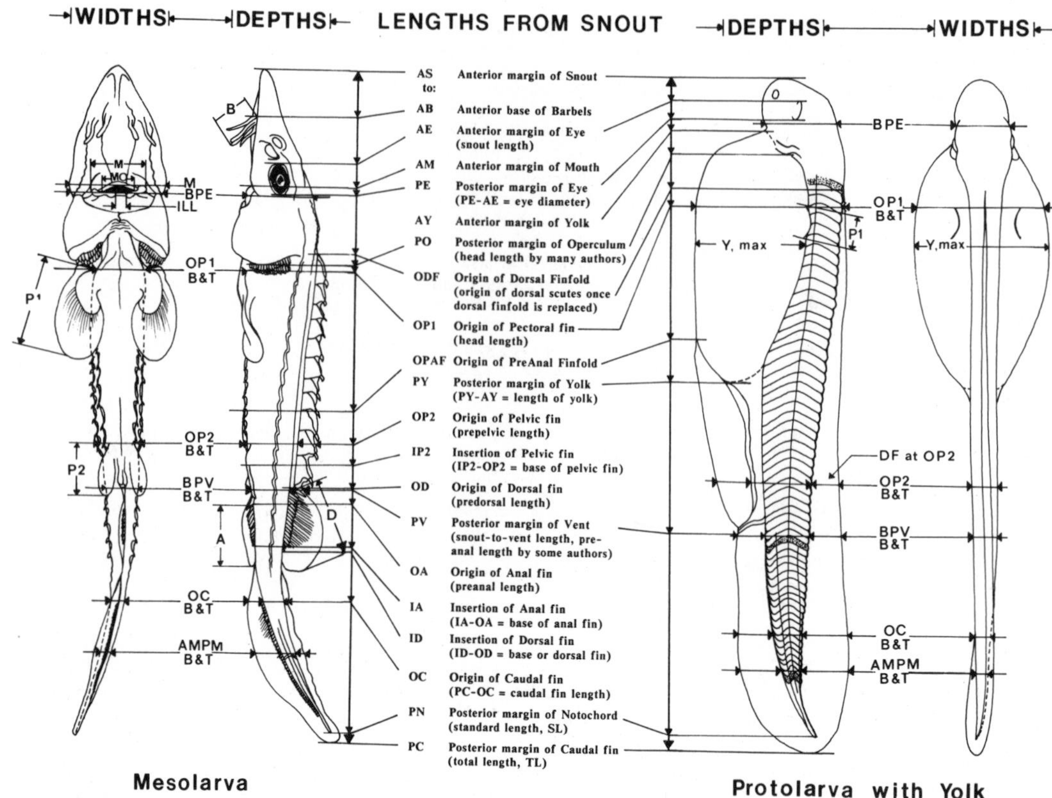

FIGURE 1.—Measurements used in morphometric analyses of sturgeon larvae. The first preanal and first and last postanal myomeres are stippled in the lateral view of the protolarva. Abbreviations not defined in the figure: **B**, barbel (length); **A**, anal fin (length); **D**, dorsal fin (length); **P1**, pectoral fin (length); **P2**, pelvic fin (length); **DF**, dorsal fin fold; **M**, mouth; **MO**, mouth opening; **ILL**, inter–lip-lobe distance; **BPE**, just behind posterior margin of eye; **BPV**, just behind posterior margin of vent; **AMPM**, anterior margin of most posterior myomere; **B&T** associated with depth and width measures, **B** for body exclusive of fin folds, fins, and scutes, and **T** for total, inclusive of these structures; **IOR**, interorbital distance (not illustrated), fleshy width between orbits of eyes. Fin lengths were measured from origin to most distal margin. Positions for OP2 and OC widths and depths prior to formation of referenced structures were approximated at ⅔ PY to PV and ½ PV to PC, respectively. Position of AMPM on some mesolarvae was approximated.

cluded all discernible rays except those incorporated in the spinelike structure along the anterior margin of pectoral fins. When practical, polarizing filters and transmitted light were used to facilitate myomere and fin ray counts. Counts also were made of the skeletal fin ray supports referred to as pterygiophores for the dorsal and anal fins and as radials for the paired fins. Finally, body scutes were counted as dorsal, lateral, and ventral (actually ventrolateral) series (DS, LS, and VS, respectively). Each series began at or near the back of the head; they ended at the anterior margin of the dorsal fin, on the lateral surface of the fleshy portion of the caudal fin, and just anterior to the pelvic fins, respectively. The number and position of scutes between fins posterior to the dorsal and ventral series were also determined. Each count was repeated at least once to ensure accuracy.

Larval phases of development are herein designated as protolarva with yolk, protolarva without apparent yolk, and mesolarva (Snyder 1976, 1983). Protolarvae have no fin rays in the median (dorsal, anal, and caudal) fins. Mesolarvae have some but an incomplete number of principal rays in the median fins or lack pelvic fins or buds. All fin rays of sturgeons are treated as principal rays, and pelvic buds develop during the protolarval phase. Some specimens categorized as protolarvae without yolk may still possess yolk, but its detection would require dissection.

Drawings were traced from enlarged photographs to provide for accurate body proportions; then the drawings were refined and completed while two or more similar-size specimens were studied under a stereozoom microscope. Final drawings may be composites of two or more specimens and are somewhat idealized (e.g., closed or frayed fins were drawn open and smooth, and curved bodies were straightened). To avoid confusion of pigmentation with other aspects of the drawings, only surface or near-surface melanophores were represented in black ink; the rest of each drawing was produced with various grades of graphite (Figures 2–9).

Results and Discussion

Specific data were recorded for 101 specimens measuring 6.2–136 mm TL. Most results for both species are similar to corresponding observations and data extracted from illustrations and descriptions by Pekovitch (1979), Taubert and Dadswell (1980), and Bath et al. (1981). The reported or suspected identity of collected specimens described or illustrated in these publications are verified by my results, except for recently hatched protolarvae. Verification of the identity of recently hatched protolarvae was inconclusive, but the results are supportive of most previously reported or suspected identities.

Morphometric Characters

Measurements for cultured and collected specimens of each species were combined for Tables 2 and 3. Data for collected protolarvae with yolk are presented separately because criteria for their identification are provisional.

In most cases, I considered morphometric characters to be diagnostically valuable if means between species were well separated and ranges were mutually exclusive, or nearly so. Statistical differences at lesser levels (e.g., 5% significance level; Student's *t*-test) are many, even between cultured and collected specimens of the same species, but most of these differences can be explained by differences in size within the developmental phase analyzed, absorption of yolk, regression of fin folds, or condition factor (depth and width measurements).

Mesolarvae and protolarvae without yolk.— Taubert and Dadswell (1980) suggested that mouth width, a diagnostic character for adult sturgeons (Ryder 1890; Vladykov and Greeley 1963), might distinguish the larvae of the two species once the mouth is formed. Bath et al. (1981) substantiated this observation for protolarvae between 14 and 21 mm TL from actual measurements on collected specimens (Table 4). However, because sample size was small and restricted to one river, and size range was very limited for specimens determined to be shortnose sturgeon, the authors called for further verification of the criteria by other researchers. Dadswell et al. (1984) cited the above papers and stated that "mouth width is the best character for separating all sizes of shortnose sturgeon and Atlantic sturgeon, including all larvae . . . ," except those bearing yolk.

The diagnostic value of mouth width is supported by my data, but, for most relationships, ranges between species do overlap (Table 4). Vladykov and Greeley (1963) considered width of the mouth opening to be a primary diagnostic character for adults. Although Bath et al. (1981) followed Vladykov and Greeley (1963) by measuring mouth width inside the lips (mouth opening, MO), their results tend to match my data better for mouth width including lips (mouth, M) than for width without lips (MO). For larvae at least up to 50 mm SL, I found width of mouth opening more difficult to measure, less consistent, and less discriminating than mouth width including lips. Some differences between my data and those of Bath et al. (1981) probably relate to their smaller sample size and use of greatest head width; I used head widths measured just behind the posterior margins of the eyes (BPE) and in line with the mouth (M). Bath et al. (1981) miscalculated their tabulated data for mouth width relative to interorbital width (a relationship used in adult keys) and actually presented the inverse, interorbital width as a percentage of mouth width (I recalculated the data for comparison in Table 4).

For larvae 15 mm to at least 50 mm SL, mouth width (hereafter the measurement includes lips) relative to head width measured at the mouth is somewhat more discriminating than mouth width relative to head length (anterior margin of snout, AS, to posterior margin of operculum, PO) and much more discriminating than mouth width relative to interorbital distance (IOR) or head width immediately behind the eyes (Table 4). Also, the character is more obvious (Figures 6, 7; Figure 7 in Dovel and Berggren 1983; Figure 4 in Dadswell et al. 1984) and easier to determine (both measurements are along the same line) than head width relative to head length. For protolarvae over 14 mm SL, mouth widths relative to head width measured at the mouth were 65% or less for Atlantic sturgeon (mean, 61%) and greater than

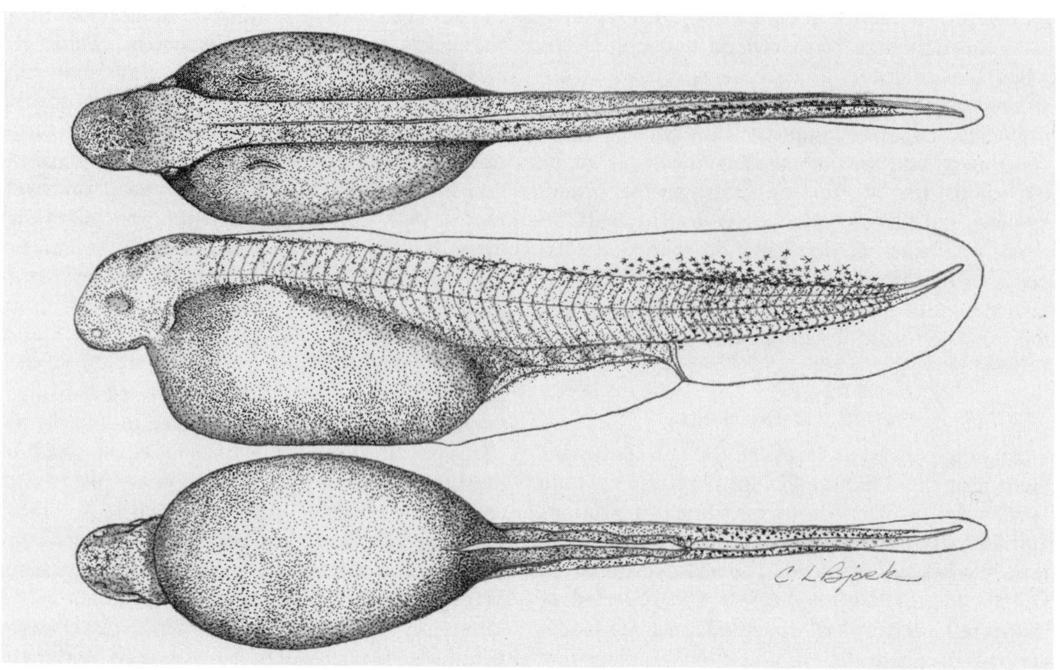

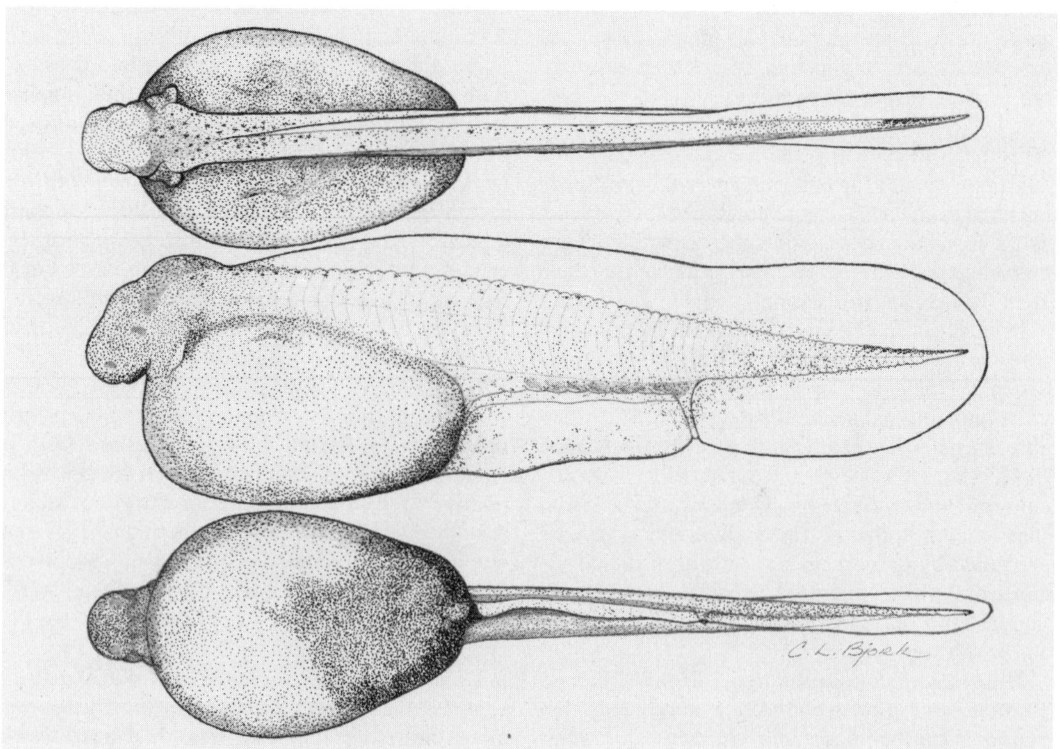

FIGURE 2.—Sturgeon protolarvae with yolk, reared at the Orangeburg National Fish Hatchery. Above: shortnose sturgeon, 10.9 mm standard length (SL), 11.1 mm total length (TL), preserved March 9, 1983, less than 12 h after hatching. Below: Atlantic sturgeon, 7.4 mm SL, 7.5 mm TL, preserved March 30, 1981, about 12 h after hatching.

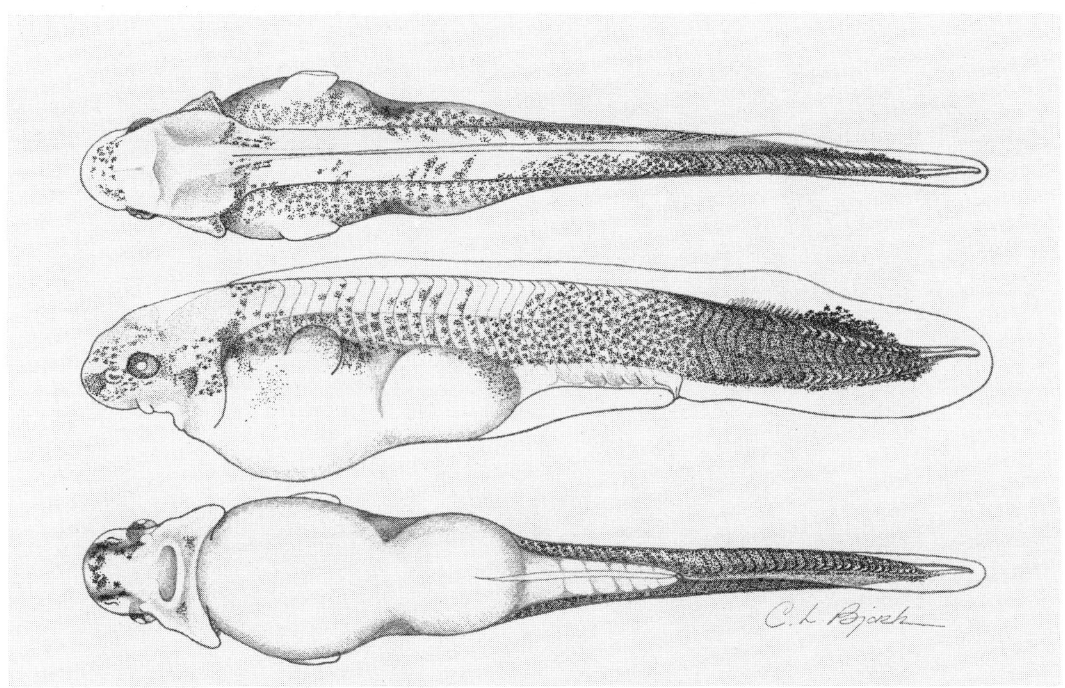

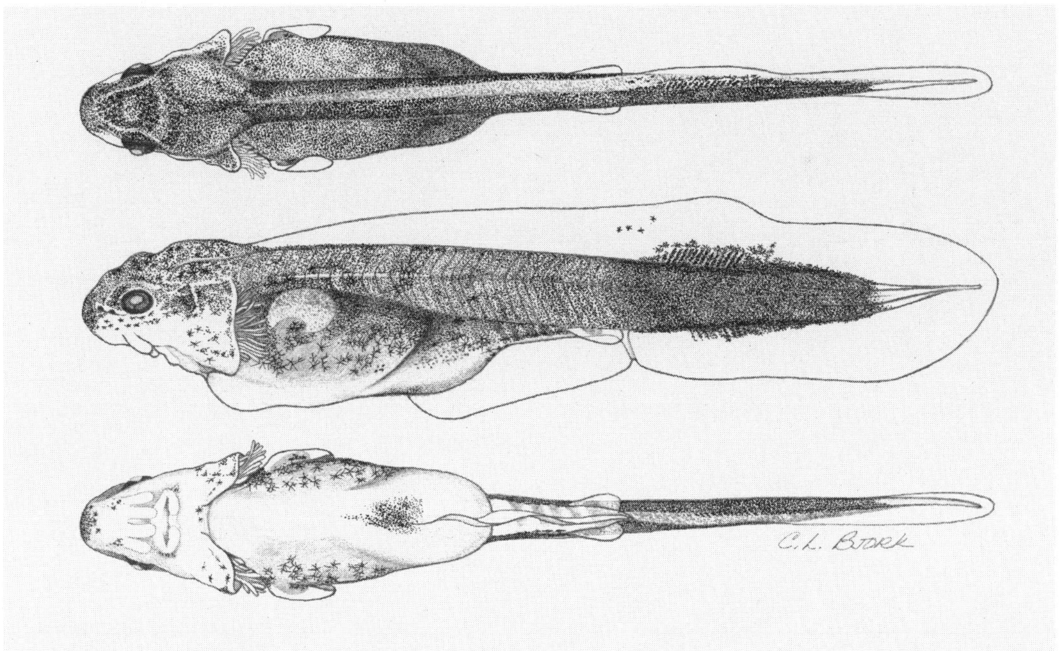

FIGURE 3.—Sturgeon protolarvae with yolk, reared at the Orangeburg National Fish Hatchery. Above: shortnose sturgeon, 12.0 mm standard length (SL), 12.1 mm total length (TL), preserved March 13, 1983, 4 d after hatching. Below: Atlantic sturgeon, 11.5 mm SL, 11.7 mm TL, preserved April 3, 1981, 4 d after hatching.

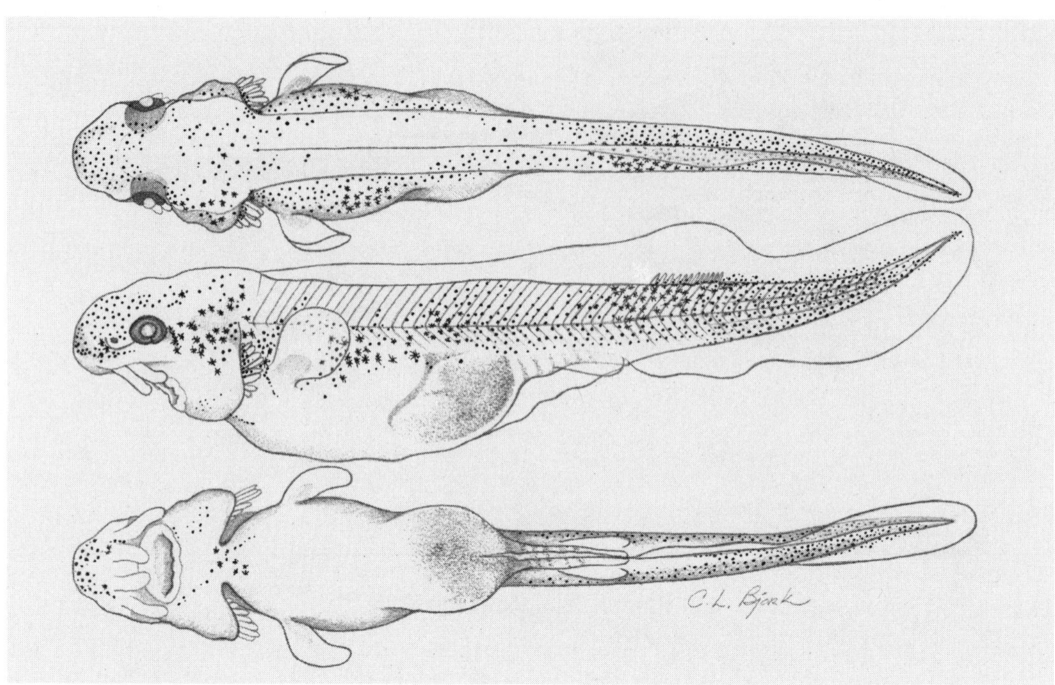

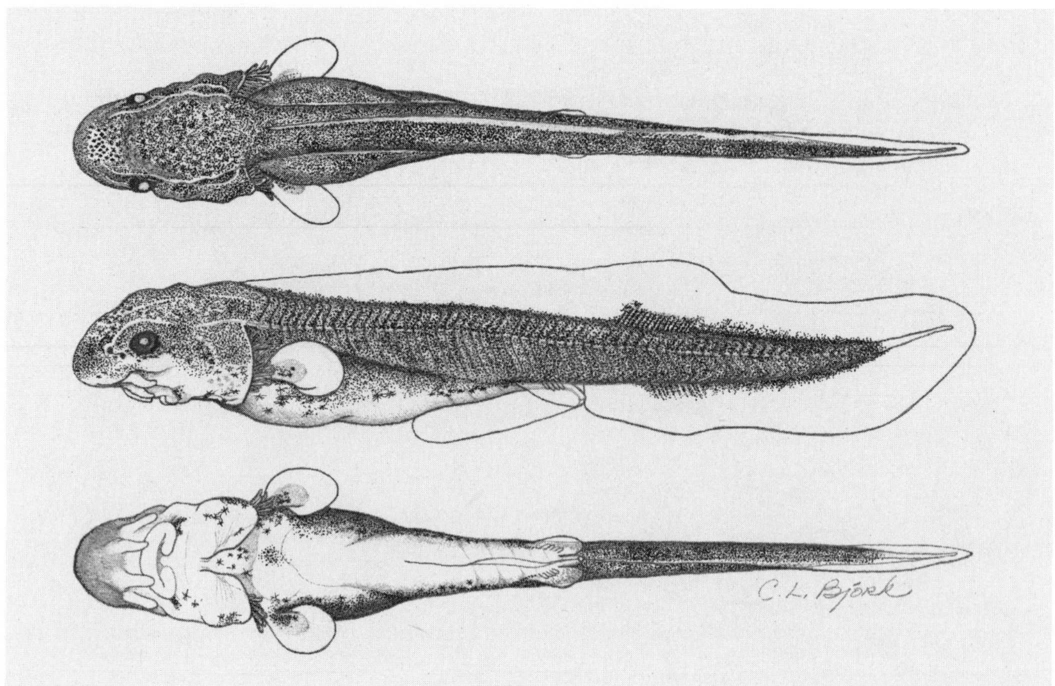

FIGURE 4.—Sturgeon protolarvae with little yolk, reared at the Orangeburg National Fish Hatchery. Above: shortnose sturgeon, 13.5 mm standard length (SL), 13.6 mm total length (TL), preserved March 15, 1983, 6 d after hatching. Below: Atlantic sturgeon, 12.9 mm SL, 13.1 mm TL, preserved April 5, 1981, 6 d after hatching.

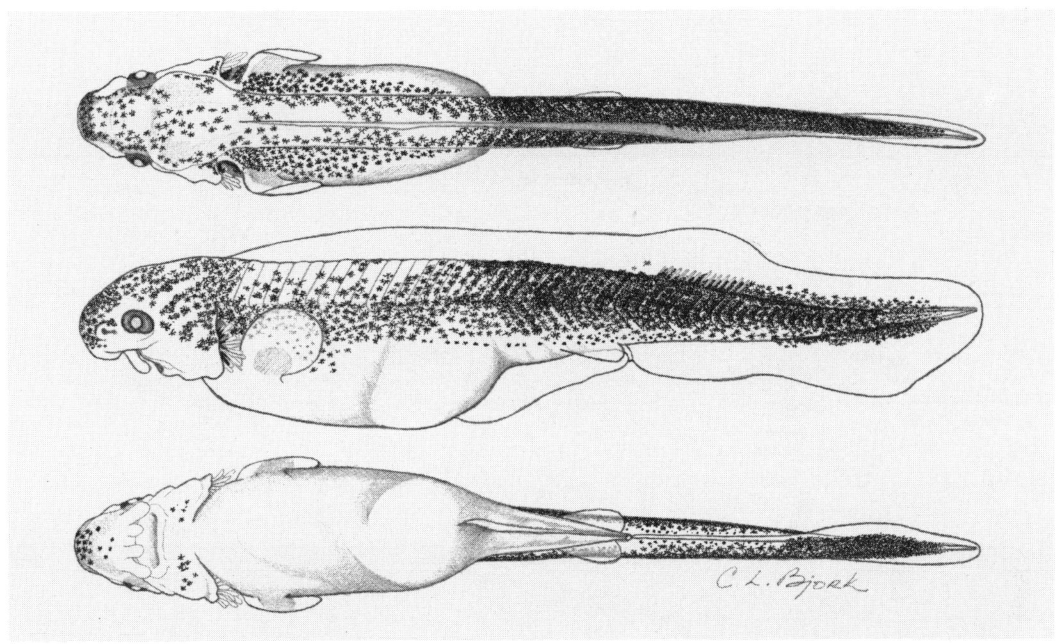

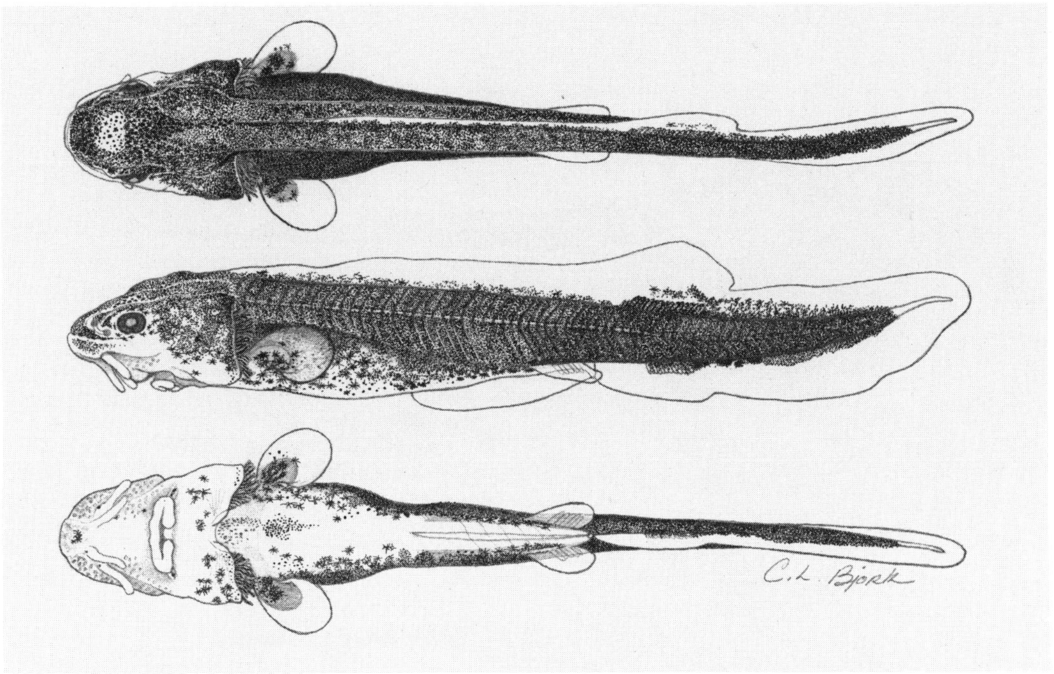

FIGURE 5.—Sturgeon protolarvae without yolk, reared at the Orangeburg National Fish Hatchery. Above: shortnose sturgeon (complete absorption of yolk is questionable), 15.1 mm standard length (SL), 15.2 mm total length (TL), preserved March 16, 1983, 7 d after hatching. Below: Atlantic sturgeon, 14.3 mm SL, 14.6 mm TL, preserved April 6, 1981, 7 d after hatching.

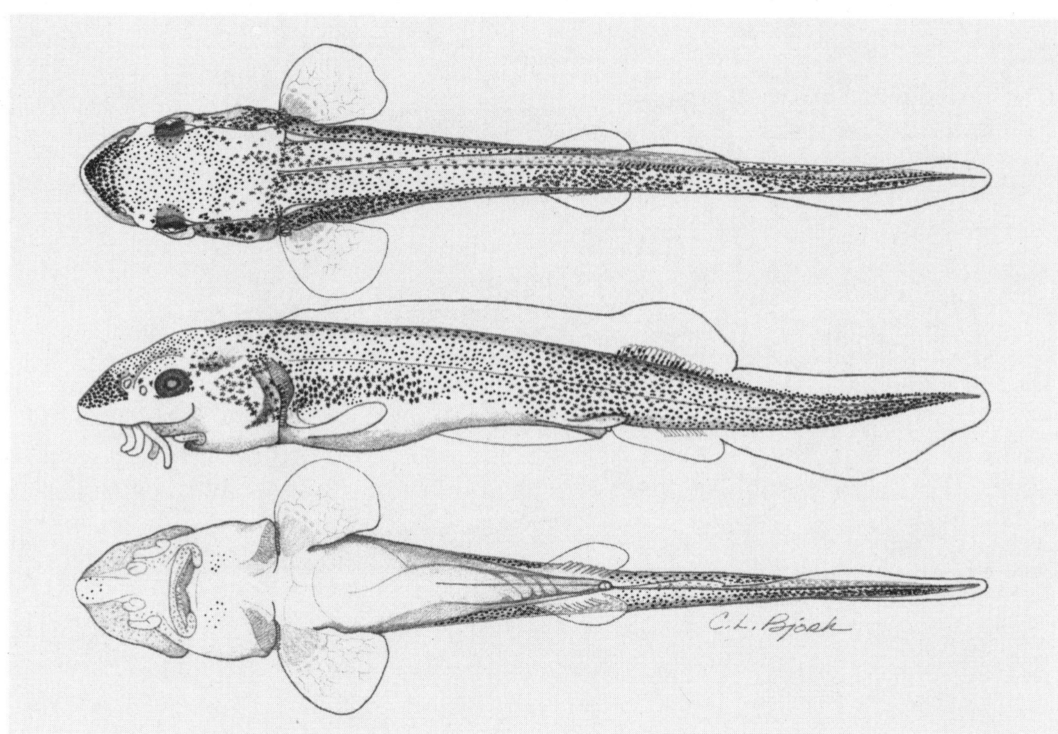

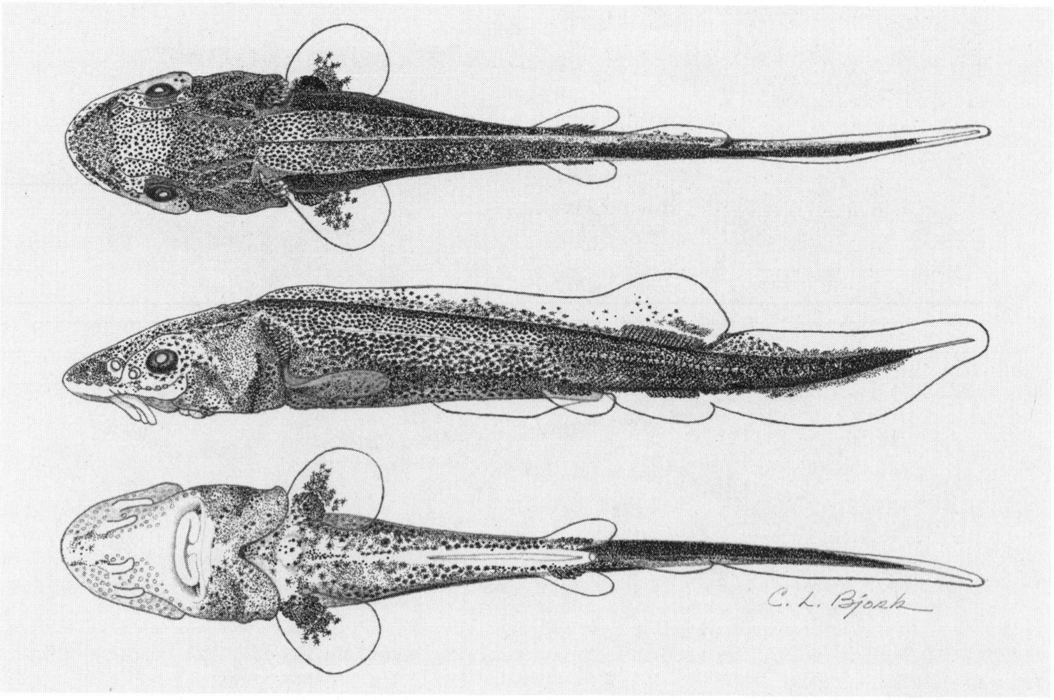

FIGURE 6.—Sturgeon protolarvae without yolk, reared at the Orangeburg National Fish Hatchery. Above: shortnose sturgeon, 17.7 mm standard length (SL), 18.0 mm total length (TL), preserved March 20, 1983, 11 d after hatching. Below: Atlantic sturgeon, 17.0 mm SL, 17.3 mm TL, preserved April 11, 1981, 12 d after hatching.

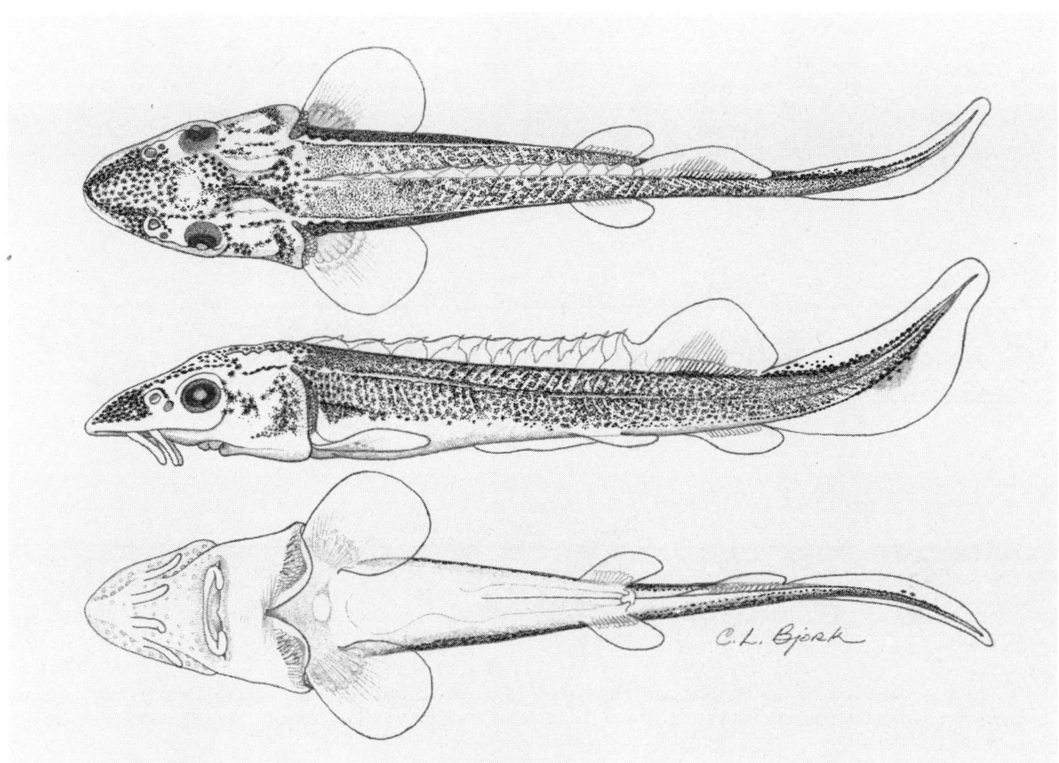

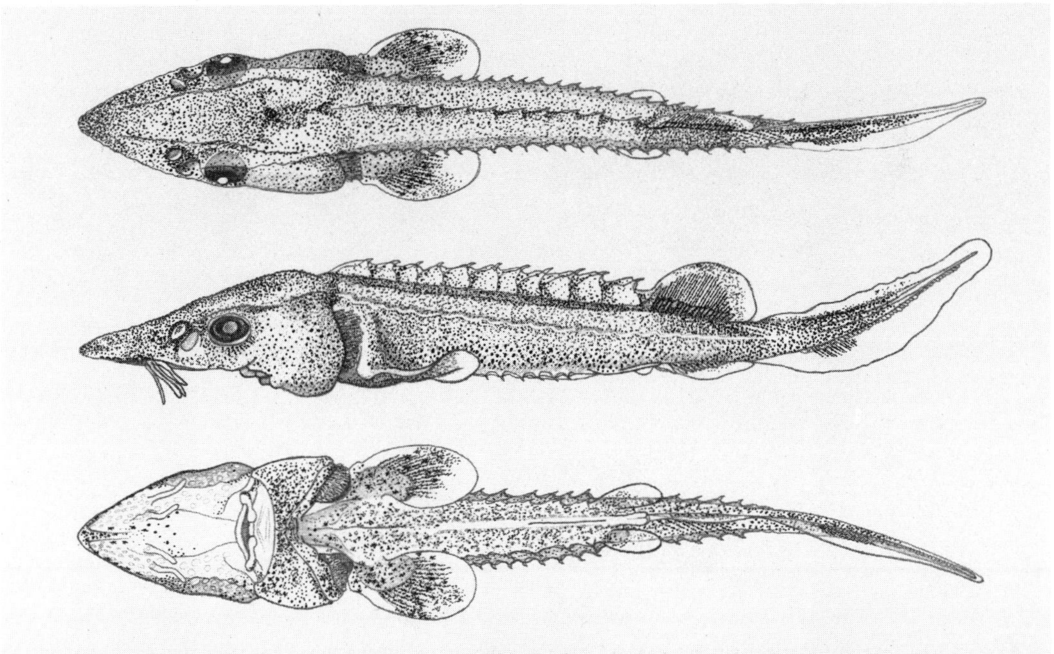

FIGURE 7.—Sturgeon mesolarvae reared at the Orangeburg National Fish Hatchery. Above, shortnose sturgeon, 24.4 mm standard length (SL), 24.9 mm total length (TL), preserved April 3, 1983, 25 d after hatching. Below: Atlantic sturgeon, 28.9 mm SL, 29.3 mm TL, preserved April 28, 1981, 29 d after hatching.

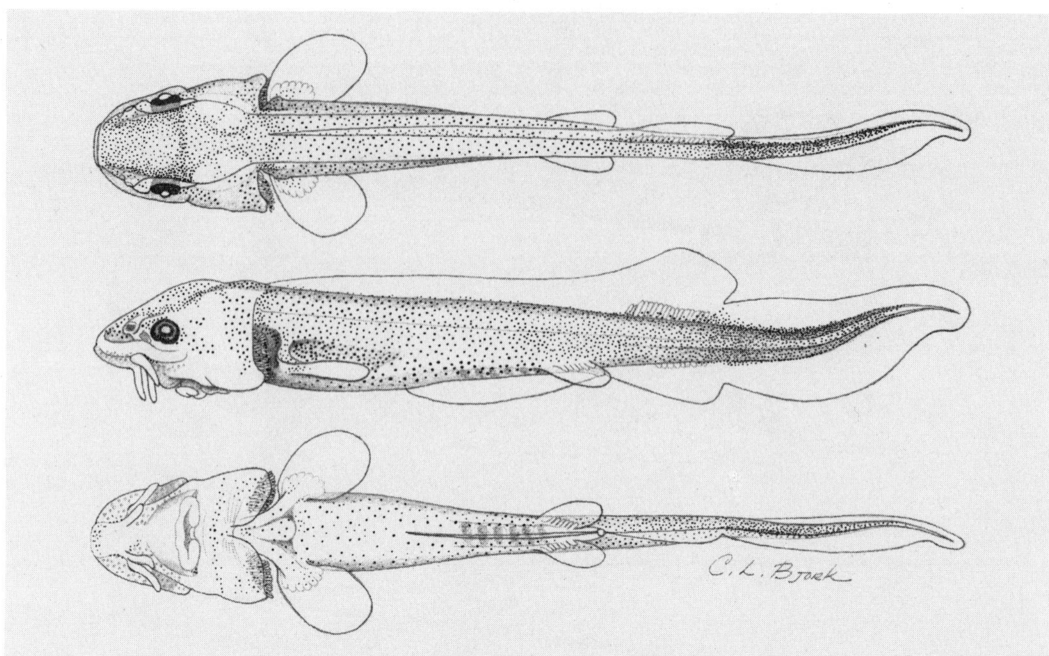

FIGURE 8.—Atlantic sturgeon protolarva without yolk, 14.0 mm standard length, 14.3 mm total length, collected and preserved from the Hudson River near Stony Point, New York (river kilometer 64), June 22, 1976 (specimen 14 in Bath et al. 1981).

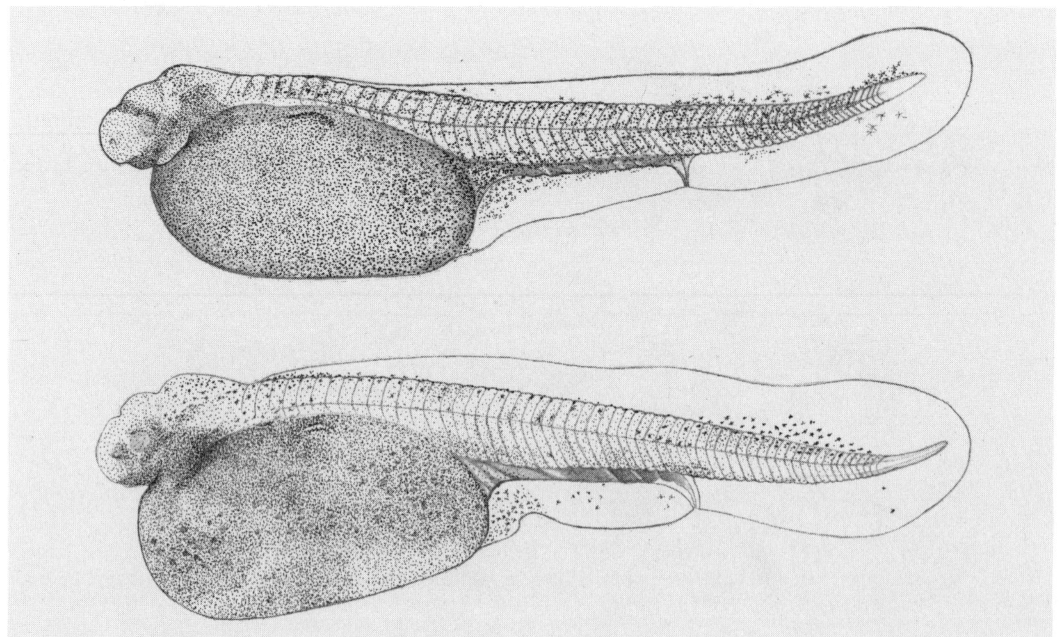

FIGURE 9.—Recently hatched sturgeon protolarvae collected 2 months apart from the same locality along the Savannah River near the U.S. Department of Energy's Savannah River Plant, South Carolina (Paller et al. 1984). Above: probable shortnose sturgeon, 9.6 mm standard length (SL), 9.7 mm total length (TL), collected and preserved March 12, 1982. Below: probable Atlantic sturgeon, 7.4 mm SL, 7.5 mm TL, collected and preserved May 21, 1982.

TABLE 2.—Summary of morphometric characters for shortnose sturgeon larvae. Protolarvae with yolk collected from the Savannah River, South Carolina and Georgia, are probable identities and, therefore, are tabulated separately from those cultured at Orangeburg National Fish Hatchery, South Carolina. Protolarvae without (obvious) yolk and mesolarvae include specimens reared at Orangeburg National Fish Hatchery and at Saint Andrews Biological Station, New Brunswick, Canada, and collected from the Hudson River, New York, the Savannah River, and Merrymeeting Bay, Maine. See Figure 1 for abbreviations and diagram of measurements. Values are rounded to nearest integer; "0" is 0 and ">0" is >0 to <0.50.

	Protolarvae, yolk						Combined sources					
	Collected			Cultured			Protolarvae, no yolk			Mesolarvae		
Measurement	Mean±SD	Range	*N*	Mean±SD	Range	*N*	Mean±SD	Range	*N*	Mean±SD	Range	*N*
SL, mm	10±1	9–11	8	12±1	10–14	9	16±2	14–21	19	42±12	24–57	9
TL, mm	10±1	9–11	8	12±1	10–14	9	17±2	15–21	19	43±12	25–58	9
Lengths, % of SL												
AS to AE	2±>0	2–3	6	4±1	2–6	9	7±1	5–9	19	13±2	10–14	9
to PE	6±>0	5–6	6	8±2	5–10	9	11±2	8–14	19	17±2	14–19	9
to AB	5±1	4–6	2	5±1	4–6	8	5±1	3–6	19	8±1	6–9	9
to AM	7±1	6–8	5	7±>0	7–8	8	10±2	7–14	19	15±2	12–17	9
to PO	11±>0	11–12	4	14±3	11–18	9	22±3	16–25	19	28±2	25–30	9
to OP1	21±1	20–23	7	20±1	19–21	9	23±3	18–27	19	28±2	25–30	9
to OP2				57±2	54–58	5	53±2	51–57	19	57±1	55–59	9
to IP2				62±2	60–64	5	58±1	57–60	19	60±1	58–61	9
to AY	8±2	6–12	8	10±4	7–18	7						
to PY	46±3	43–50	8	48±2	47–51	7						
to ODF	20±3	17–26	8	18±1	17–20	9	21±3	16–29	19	25±2	22–27	9
to OPAF	42±3	39–46	7	44±3	39–48	9	41±3	35–46	19	44±1	43–46	7
to PV	69±3	63–74	8	65±2	61–69	9	60±1	58–62	19	62±1	60–64	9
to OD				63	63	1	62±2	58–65	19	64±2	62–67	9
to ID							71±2	69–75	13	73±2	71–75	9
to OA							66±2	64–69	11	68±2	65–70	9
to IA							72±2	70–75	11	74±1	72–75	9
to OC										78±2	74–81	8
to PC	102±>0	101–102	8	101±>0	100–2	9	102±1	101–3	19	102±1	100–2	9
B	>0±>0	0–1	8	2±1	0–4	9	5±2	1–8	19	6±1	5–7	9
P1	2±2	1–6	7	7±2	5–11	9	13±2	9–15	19	14±1	13–15	9
P2	0±0	0–0	8	3±3	0–7	9	8±2	4–11	19	9±1	8–10	9
D							13±1	13–14	10	11±2	9–15	9
A							9±1	7–10	7	8±1	7–10	9
Depths, % of SL												
at BPE	10±1	8–11	7	10±1	9–12	9	11±1	9–13	19	10±1	9–12	9
at OP1,B	21±3	16–25	6	20±2	17–23	9	13±2	10–17	19	12±1	11–13	9
at OP1,T	21±3	17–25	5	21±2	17–24	9	14±2	12–18	19	13±1	13–14	9
at OP2,B	8±>0	7–8	8	10±1	9–11	9	8±1	7–10	19	9±1	7–9	9
at OP2,T	14±2	12–18	8	16±3	11–20	9	13±2	11–17	18	12±1	11–13	9
at BPV,B	5±>0	5–6	8	7±1	7–8	9	7±1	5–8	19	7±>0	7–8	9
at BPV,T	13±2	10–16	8	13±2	11–17	9	12±1	9–13	18	10±1	9–12	9
at OC,B	4±>0	3–5	7	4±1	4–6	9	4±1	3–5	19	4±1	3–5	9
at OC,T	14±2	12–18	7	14±2	12–17	9	10±1	8–12	17	5±2	4–10	9
at AMPM,B	2±>0	2–3	8	2±>0	2–3	9	2±>0	2–2	19	2±>0	1–2	9
at AMPM,T	14±2	11–17	8	14±2	11–17	9	10±1	9–13	18	7±1	5–8	9
DF at OP2	3±1	2–4	8	3±1	1–4	9						
Yolk, maximum	18±5	11–22	4	17±2	14–19	7						
Widths, % of SL												
at BPE	8±1	7–8	8	10±1	8–12	9	14±2	10–19	19	15±1	14–16	9
at M	20±5	13–25	7	10±2	8–12	8	14±2	10–17	19	14±1	13–16	9
at OP1	4±>0	3–5	8	16±4	11–21	9	11±1	9–13	19	14±1	11–17	9
at OP2,B				6±1	5–6	9	5±>0	4–6	19	6±1	6–7	9
at OP2,T	4±>0	3–4	8							9±1	8–10	8
at BPV,B				5±1	4–6	9	4±>0	4–5	19	5±>0	5–5	9
at BPV,T	3±>0	2–3	8							6±1	5–7	8
at OC	2±>0	2–2	8	4±1	3–5	9	3±>0	2–3	19	2±1	2–3	9
at AMPM				2±>0	1–2	9	1±>0	1–1	19	1±>0	1–1	9
M				8±1	8–9	5	10±1	8–14	18	10±1	9–11	9
MO				6±1	5–7	5	7±1	5–10	18	7±>0	6–7	9
IOR	4±>0	4–5	4	7±2	4–10	8	9±1	8–12	18	10±1	8–11	9
ILL				5	5	1	4±>0	4–5	14	3±1	2–5	9
Yolk, maximum	23±3	18–25	5	18±3	14–21	7						

TABLE 3.—Summary of morphometric characters for Atlantic sturgeon larvae. Protolarvae with yolk collected from the Savannah River, South Carolina and Georgia, are probable identities and, therefore, are tabulated separately from those cultured at Orangeburg National Fish Hatchery, South Carolina. Protolarvae without (obvious) yolk and mesolarvae include specimens reared at Orangeburg National Fish Hatchery and collected from Hudson River, New York, Savannah River, and Winyah Bay, South Carolina. See Figure 1 for abbreviations and diagram of measurements. Values are rounded to nearest integer; "0" is 0 and ">0" is >0 to <0.50.

	Protolarvae, yolk						Combined sources					
	Collected			Cultured			Protolarvae, no yolk			Mesolarvae		
Measurement	Mean±SD	Range	*N*	Mean±SD	Range	*N*	Mean±SD	Range	*N*	Mean±SD	Range	*N*
SL, mm	8±1	6–9	6	11±2	9–13	9	16±2	14–21	15	37±9	29–47	6
TL, mm	8±1	6–9	6	11±2	9–13	9	16±2	14–21	15	38±9	29–47	6
Lengths, % of SL												
AS to AE	3±1	3–4	5	4±1	3–5	8	7±2	4–10	15	14±2	12–16	6
to PE	7±>0	7–7	5	8±1	7–9	8	11±2	8–14	15	18±2	16–20	6
to AB	4±1	3–6	5	4±1	4–5	9	4±1	3–7	15	8±2	7–11	6
to AM	7±1	6–8	5	7±1	6–9	9	11±2	7–13	15	16±3	14–19	6
to PO	13±1	12–14	5	15±3	11–20	9	23±2	20–26	15	29±2	27–31	6
to OP1	24±3	21–27	5	20±2	17–24	9	22±2	20–26	15	30±2	27–32	6
to OP2				50±2	48–52	5	52±2	49–59	15	57±1	55–58	6
to IP2				58±2	55–60	5	57±2	55–62	15	60±1	59–61	6
to AY	8±1	7–10	6	12±5	6–21	9						
to PY	48±1	47–50	6	43±5	34–47	9						
to ODF	20±2	19–23	6	17±1	15–19	9	20±2	17–27	15	27±2	24–29	6
to OPAF	45±3	41–49	6	39±3	34–42	9	39±3	35–45	15	46±4	43–52	6
to PV	73±1	72–74	6	63±6	56–71	9	59±2	56–64	15	62±1	61–64	6
to OD				60±2	58–62	3	60±2	57–64	15	63±2	61–65	6
to ID							70±2	67–74	15	72±1	70–73	6
to OA				64±2	62–66	3	64±2	61–70	15	67±2	65–68	6
to IA							70±2	67–75	14	72±1	70–73	6
to OC										77±1	75–78	5
to PC	102±1	101–103	6	102±>0	101–102	9	101±1	101–102	15	101±1	100–101	6
B	1±>0	0–1	6	3±2	1–5	9	5±1	4–6	15	5±1	5–6	6
P1	5±1	4–6	5	7±2	5–10	9	12±2	10–15	15	13±1	12–14	6
P2	0±0	0–0	6	5±4	0–9	9	8±1	7–9	15	8±1	7–9	6
D							13±1	12–15	11	12±1	10–14	6
A							8±1	7–9	10	8±1	7–10	6
Depths, % of SL												
at BPE	11±2	10–14	6	10±1	9–11	9	11±1	10–14	15	11±1	9–12	6
at OP1,B	24±4	19–29	6	19±6	11–26	9	12±1	10–15	15	12±1	11–12	6
at OP1,T	25±4	20–30	5	20±7	12–27	9	13±1	11–16	15	13±1	12–14	6
at OP2,B	8±>0	7–8	6	9±1	8–9	9	8±1	7–10	15	8±2	5–9	6
at OP2,T	15±2	11–18	6	20±2	17–23	7	14±2	11–18	15	11±2	10–14	6
at BPV,B	5±1	5–6	6	7±1	6–7	9	7±1	6–8	15	6±1	6–8	6
at BPV,T	14±1	13–15	5	16±2	14–19	9	12±1	10–16	15	9±2	7–11	6
at OC,B	4±>0	4–5	6	4±1	3–5	9	4±0	3–5	15	3±>0	3–4	6
at OC,T	14±1	13–16	5	18±3	13–22	8	11±1	9–14	14	6±1	5–9	6
at AMPM,B	2±>0	2–3	6	2±>0	2–3	9	2±0	2–3	15	2±>0	1–2	6
at AMPM,T	14±2	11–16	6	16±1	13–18	9	10±2	8–14	14	7±1	6–8	6
DF at OP2	3±1	2–4	6	5±1	4–6	9						
Yolk, maximum	20±3	17–24	6	16±6	8–22	8						
Widths, % of SL												
at BPE	8±>0	8–9	6	9±1	7–12	9	13±2	10–16	15	13±1	12–14	6
at M				10±1	9–12	5	13±1	12–15	15	13±>0	13–14	6
at OP1	22±3	18–26	5	16±6	9–23	9	10±2	9–17	15	14±1	13–16	6
at OP2,B	4±1	3–5	6	4±1	3–5	9	4±1	3–6	15	5±>0	4–5	6
at OP2,T										7±1	6–8	6
at BPV,B	4±1	3–4	6	4±1	3–4	9	4±1	3–6	15	4±>0	4–5	6
at BPV,T										6±1	5–8	6
at OC	3±1	2–4	6	2±1	2–3	9	3±1	2–5	14	2±1	1–3	6
at AMPM	2±>0	2–2	6	2±>0	1–2	9	1±0	1–2	14	1±>0	1–1	6
M				7±>0	6–8	5	8±0	8–9	15	7±>0	7–8	6
MO				5±1	4–6	5	6±0	5–7	15	5±>0	5–6	6
IOR	4±1	3–5	3	6±1	4–8	8	9±1	8–10	15	9±1	9–10	6
ILL				1±>0	1–2	2	1±0	1–1	14	1±>0	1–1	5
Yolk, maximum	23±3	19–26	5	18±5	12–23	8						

TABLE 4.—Inter–lip-lobe distance, mouth width (with lips), and mouth-opening width as a percentage of selected head measurements for shortnose sturgeon and Atlantic sturgeon larvae without yolk and greater than 13 mm standard (SL) or total (TL) length. Data are based on both cultured and collected specimens. Data from Bath et al. (1981) for mouth-opening width are provided for comparison. Where the number and size range of specimens are adequate, the degree of difference between species means and ranges is indicative of a character's diagnostic value. Negative range differences represent overlap; asterisk (*) indicates one range is completely included within the other. Inter–lip-lobe distance (ILL) is width between lobes of lower lip; mouth width (M) includes lips; mouth-opening width (MO) excludes lips; head width was measured immediately behind eyes (W at BPE), at mouth (W at M), and at greatest width (GHW); head length was measured to posterior margin of opercula (AS to PO); interorbital distance (IOR) is least width between fleshy orbits of eyes.

Measurement	Shortnose sturgeon			Atlantic sturgeon			Difference	
	Mean±SD	Range	*N*	Mean±SD	Range	*N*	Mean	Range
Inter–lip-lobe distance								
Protolarvae >13 mmSL								
mm SL		14–21			14–21			
% M	46±7	34–58	14	14±3	6–18	14	32	16
% W at M	34±6	27–47	14	9±2	3–13	14	25	15
% W at BPE	33±6	25–47	14	9±2	3–14	14	24	11
% AS to PO	21±5	15–29	14	5±1	2–7	14	16	8
Mesolarvae								
mm SL		24–57			30–47			
% M	32±5	26–42	9	15±3	10–18	5	17	8
% W at M	22±4	17–31	9	8±2	6–10	5	14	7
% W at BPE	22±4	16–31	9	8±2	6–11	5	14	5
% AS to PO	11±3	9–18	9	4±1	2–5	5	7	4
Mouth width (with lips)								
Protolarvae >13 mm SL								
mm SL		14–21			14–21			
% W at M	75±5	66–88	18	63±4	57–72	15	12	−6
% W at BPE	75±9	59–100	18	64±6	55–79	15	11	−20
% AS to PO	47±4	38–55	18	38±3	33–42	15	9	−4
% IOR	107±11	87–140	17	93±7	79–105	15	14	−18
Protolarvae >14 mm SL								
mm SL		15–21			15–21			
% W at M	75±6	66–88	17	61±3	57–65	9	14	1
% W at BPE	75±9	59–100	17	60±4	55–68	9	15	−9
% AS to PO	46±4	38–55	17	36±3	33–42	9	10	−4
% IOR	107±11	87–140	16	92±4	87–100	9	15	−13*
Mesolarvae								
mm SL		24–57			29–47			
% W at M	69±4	62–74	9	57±1	56–59	6	12	3
% W at BPE	67±4	61–74	9	56±3	53–61	6	11	0
% AS to PO	35±3	31–43	9	26±2	24–28	6	9	3
% IOR	101±4	84–119	9	82±2	73–88	6	19	−4
Mouth-opening width								
Protolarvae >13 mm SL								
mm SL		14–21			14–21			
% W at M	50±6	42–67	18	44±4	37–50	15	6	−8
% W at BPE	51±9	40–76	18	45±5	35–52	15	6	−12
% AS to PO	31±4	25–42	18	26±3	22–31	15	5	−6
% IOR	72±11	57–107	17	65±6	57–75	15	7	−18
Protolarvae >14 mm SL								
mm SL		15–21			15–21			
% W at M	50±6	42–67	17	43±5	27–50	9	7	−8
% W at BPE	50±9	40–76	17	43±5	35–50	9	7	−10
% AS to PO	31±4	25–42	17	26±3	22–31	9	5	−6
% IOR	72±11	57–107	16	65±5	60–73	9	7	−13*
Mesolarvae								
mm SL		24–57			29–47			
% W at M	47±2	43–50	9	41±2	39–44	6	6	−1
% W at BPE	46±4	42–51	9	40±2	39–43	6	6	−1
% AS to PO	24±2	21–30	9	19±2	16–21	6	5	0
%IOR	70±7	61–83	9	58±2	53–64	6	12	−3
Mouth-opening width—from Bath et al. (1981)								
Protolarvae >13 mm TL								
mm TL		15–16			14–21			
% GWH	73±5	65–75	4	52±3	48–55	5	21	10
% AS to PO	44±2	43–47	4	36±3	33–41	7	8	2
% IOR	147±7	136–150	4	130±23	100–157	5	17	−14*
Protolarvae >14 mm TL								
mm TL		15–16			17–21			
% GHW	73±5	65–75	4	51±3	48–55	4	22	10
% AS to PO	44±2	43–47	4	34±1	33–36	4	10	7
% IOR	147±7	136–150	4	111±15	100–121	2	36	15
Mesolarvae								
mm TL					32–37			
% GHW				48		1		
% AS to PO				26±2	25–27	2		
% IOR				97±21	82–112	2		

65% for shortnose sturgeon (mean, 75%). However, the boundary between values for the two species (65–66%) might not hold in all cases because inclusion of 14-mm-SL specimens in the data set extends the range for Atlantic sturgeon up to 72%. For mesolarvae, mouth widths relative to head width measured at the mouth were less than 60% for Atlantic sturgeon (mean of 57%) and greater than 61% for shortnose sturgeon (mean of 69%).

The distance between lobes of the lower lip (inter–lip-lobe distance, ILL) is the most diagnostically valuable morphometric character considered in this study (Tables 2–4). The character was illustrated by Dovel (1979), noted by Bath et al. (1981) as a personal communication from Dovel, and specified by Dovel and Berggren (1983) as a better distinguishing character than mouth width.

Although diagnostically very strong regardless of the head dimension it is related to, differences in inter–lip-lobe distance are particularly obvious when compared to mouth width, even without actual measurement (Table 4; Figures 6–8; Figure 7 in Dovel and Berggren 1983; Figure 4 in Dadswell et al. 1984). Dovel and Berggren (1983) observed that for Atlantic sturgeon larvae this distance is usually "less than a third of the total width of the mouth." I found that this distance was less than 19% of mouth width for Atlantic sturgeon (means of 14% for protolarvae longer than 13 mm SL and 15% for mesolarvae) and greater than 25% for shortnose sturgeon (means of 46% and 32%, respectively).

Dovel and Berggren (1983) stated that there are differences between Atlantic and shortnose sturgeon larvae in the size and shape of their barbels. I observed no obvious differences in shape or consistent differences in size relative to developmental state (length B in Tables 2, 3)

Protolarvae with little yolk.—Protolarvae approaching yolk depletion (12–13 mm, rarely 11 mm SL) sometimes are distinguishable on the basis of mouth width and inter–lip-lobe distance criteria discussed. However, these characters must be used cautiously and, if possible, in combination with other criteria.

Position of recently formed pelvic fin buds also might have diagnostic value for larvae approaching yolk depletion. The origin of the bud (AS to origin of pelvic fin, OP2) was greater than 53% SL for shortnose sturgeon and equal to or less than 53% for Atlantic sturgeon protolarvae with yolk (Tables 2, 3). However, all measurements were from cultured specimens; verification with collected material is needed. Also, pelvic fin positions in protolarvae without yolk are nearly identical (Tables 2, 3), and in a few instances, the between-species relationship is reversed—48–51% of TL for shortnose sturgeon and 51–55% of TL for Atlantic sturgeon based on data from Pekovitch (1979) and illustrations from Bath et al. (1981).

Protolarvae with yolk.—Morphometric characters are of limited value in identifying recently hatched protolarvae of these sturgeons. The most obvious and consistent difference, dorsal fin-fold depth, is reflected somewhat in total depths, especially at the origin of future or recently formed pelvic fin buds for cultured specimens (OP2,T and DF at OP2, Tables 2, 3) and origin of pectoral fin buds for collected specimens (OP1,T). The anterior portion of the dorsal fin fold tends to be much shallower in yolk-bearing protolarvae of shortnose sturgeon than in Atlantic sturgeon (Figures 2, 9). In association with this difference, the anterior portion of the dorsal fin fold often tends to be slightly concave in shortnose sturgeon and convex in Atlantic sturgeon. For cultured specimens examined from hatching through 13 mm SL, dorsal fin-fold depth at origin of the future or recently formed pelvic fin buds (DF at OP2) is always 36% of body depth (OP2,B) or less for shortnose sturgeon and 45% or greater for Atlantic sturgeon (Figure 10). However, addition of corresponding data for collected protolarvae with yolk (believed to be accurately identified based on other criteria) obscures this distinction for specimens with fin-fold depths over 35% of body depth. Nearly all collected Atlantic sturgeon have intermediate fin-fold depths between 36 and 50% of body depth at the pelvic fin bud origin, and fin-fold depths for collected shortnose sturgeon varied between 22 and 56% of body depth.

Meristic Characters

Myomere counts were nearly identical for the two species and accordingly have no diagnostic value. Means for both species examined for this study were 38 preanal, 22 or 23 postanal, and 60 or 61 total myomeres (Table 5). Mean preanal and total myomere counts reported by Pekovitch (1979) for shortnose sturgeon were one to two units less, those reported by Taubert and Dadswell (1980) for shortnose sturgeon were four to five units less, and those reported by Bath et al. (1981) for Atlantic sturgeon were one to four units less. For specimens examined in this study, larvae from the Saint John and Hudson rivers have myomere counts similar to

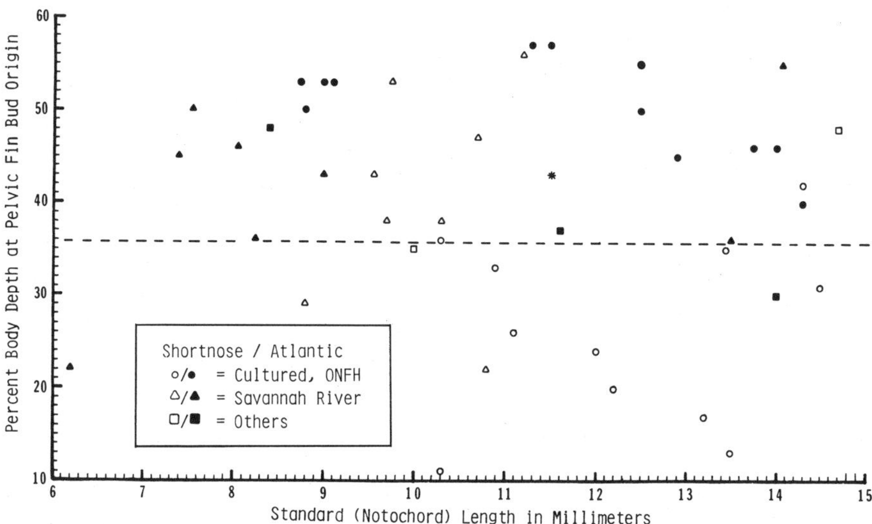

FIGURE 10.—Dorsal fin-fold depth as a percentage of body depth (fin folds excluded) at the origin (or approximated origin) of the pelvic fin buds for shortnose sturgeon and Atlantic sturgeon protolarvae before and shortly after complete yolk absorption. "Other" data were extracted from photographs in Taubert and Dadswell (1980) and Bath et al. (1981). The asterisk represents a datum extracted from a drawing reported to be of Atlantic sturgeon by Ryder (1890). Between 7 and 13 mm standard length, all data beneath the broken horizontal line represent only shortnose sturgeon; ONFH is Orangeburg National Fish Hatchery.

those from South Carolina populations. Accordingly, the discrepancies among authors probably reflect differences in criteria or technique for myomere counts rather than differences between species or geographically segregated populations. The most anterior and posterior myomeres are often difficult to discern and may be overlooked; some reported counts may exclude all anterior segments that are strictly epaxial or dorsal to the horizontal myoseptum. The number of myomeres illustrated for a recently hatched Atlantic sturgeon in Ryder (1890; 61 preanal, 35 postanal, and 96 total) is inaccurate.

Pelvic and anal fin ray counts effectively differentiate specimens larger than 50–60 mm SL (Table 6). Pelvic fin ray counts are discrete: 17–22 pelvic rays for shortnose sturgeon versus 26–35 for Atlantic sturgeon. In contrast, the diagnostic value of anal fin ray counts is limited by range overlap: 18–24 for shortnose sturgeon versus 22–32 for Atlantic sturgeon. These counts are based on original and published data and include rare or questionable observations.

Published reports of caudal fin ray counts, 60 for shortnose sturgeon and 90 for Atlantic stur-

TABLE 5.—Myomere counts for cultured and collected larvae of shortnose sturgeon and Atlantic sturgeon.

	Shortnose sturgeon			Atlantic sturgeon		
Myomere count	Mean±SD	Range	N	Mean±SD	Range	N
			Cultured			
Preanal	38±2	35–41	24	38±1	36–40	20
Postanal	22±2	20–26	22	23±1	22–24	19
Total	60±2	58–65	21	61±1	59–62	19
			Collected			
Preanal	38±1	36–40	13	38±1	36–40	13
Postanal	24±2	19–25	8	21±2	18–24	10
Total	62±2	58–65	8	59±2	56–61	10
			Combined			
Preanal	38±2	35–41	37	38±1	36–40	33
Postanal	23±2	19–26	30	22±2	18–24	29
Total	61±2	58–65	29	60±2	59–62	29

TABLE 6.—Selected meristic characters for shortnose sturgeon and Atlantic sturgeon as summarized from original observations and the literature. Original counts are believed to represent adult complements. Where appropriate and notable, mean or model values are underlined. Rare or questionable extremes are enclosed in parentheses.

	Shortnose sturgeon		Atlantic sturgeon	
Character	Count	Source[a]	Count	Source[a]
Fin rays				
Dorsal				
Original	32–34	2, 57–57	34–41	12, 39–136
Literature	33–42	A–C, F–I	30–46	A–C, E–H
Anal				
Original	18	1, 57	24–27	11, 47–136
Literature	18–19–22–24	A–D, F–I	(22)23–30(32)	A–H
Caudal				
Original			92[b]	1, 136
Literature	60	A, H	90	A
Pectoral[c]				
Original	25–27	6, 36–57	(27)28–36(38)	12, 39–136
Literature	30–31	A, H	35–41	A, H
Pelvic				
Original	19–22	3, 51–57	28–33(35)	9, 58–136
Literature	17–21	A, H	26–29	A, H
Fin ray supports				
Dorsal				
Original	17–19(23)	28, 14–57	15–16–17–18	33, 11–107
Literature	15–17	J	13	A
Anal				
Original	10–12(13)	13, 16–57	10–11–12–13	30, 13–136
Literature	10	J	10	A
Caudal				
Original	23–29	8, 34–57	(29)31–43	11, 29–107
Literature			25	A
Pectoral				
Original	7–8	25, 15–57	7–8(9)	20, 13–58
Pelvic				
Original	(7)8–9	22, 17–57	7–8–9(10)	28, 13–107
Literature	6–8	J		
Scutes series[d]				
Dorsal				
Original	10–11–14	8, 34–57	9–11–12	12, 39–136
Literature	(7)8–10–11–13	A–I	(7)10–11–14(16)	A–H
Lateral				
Original	24–31	3, 51–57	(23)24–27	12, 39–136
Literature	(21)22–25–32–34	A–I	24–26–29–35(36)	A–H
Ventral				
Original	6–7–8–9	8, 34–57	8–9–10(11)	12, 39–136
Literature	(3)6–7–9–11	A–I	8–9–11–12(14)	A–D, F–H
Gill rakers				
First arch				
Literature	22–25–28–32	E, G–I	15–22–27	E, G–I

[a]Number of specimens and size range (mm standard length) for original counts; literature sources: A, Ryder (1890); B, Jordan and Evermann (1896); C, Hildebrand and Schroeder (1928); D, Bigelow and Schroeder (1953); E, Vladykov and Greeley (1963); F, Moore (1968); G, Scott and Crossman (1973); H, Jones et al. (1978); I, Dadswell et al. (1984); and J, Pekovitch (1979, based on four specimens 16.3–18.2 mm total length).

[b]About 23 rays in lower lobe, 58 in upper lobe, and 11 at tip, which is separated from upper lobe by a narrow constriction; count at tip might not be complete.

[c]Original counts do not include rays incorporated in spinelike structure along anterior margin of fin.

[d]At least for original counts, dorsal series of scutes includes modified scute at anterior margin of dorsal fin. Original counts and most previously published counts of ventral or ventrolateral series end with a scute just anterior to pelvic fin. Some published counts might include paired scutes between vent and anal fin.

geon, indicate that this character also is discriminating (Table 6). However, acquisition of the adult count of caudal fin rays is the last criterion for transition to the juvenile period in sturgeons, so the character cannot be applied until very late in the larval period. Also, adult counts need further verification based on larger specimens.

Caudal fin ray supports, 23–29 for shortnose sturgeon and 29–43 for Atlantic sturgeon, might be diagnostic even for larvae as small as 29 mm SL (Table 6). However, Ryder (1890) reported 25 caudal fin ray supports for Atlantic sturgeon. If the latter observation is accurate, the diagnostic value of the character must, at least for the present, be limited to counts well over 29, which would represent Atlantic sturgeon.

The number of scutes (bony shields and plates) in major series or rows is similar for both species (Table 6). However, other more posteriorly located scutes differ in number and position (Scott and Crossman 1973). Atlantic sturgeon larvae 50 mm or larger are characterized by a ridge along each side of both the dorsal midline, between the dorsal and caudal fins, and one along the ventral midline, between the anal and caudal fins. The dorsal ridges are replaced in larger larvae by (typically) three scutes on each side. Between the anal and caudal fins, the ridges are replaced by one or two scutes on each side followed by an elongate median scute that extends onto the anterior margin of the caudal fin. In contrast, 51-mm-SL shortnose sturgeons have single median ridges posterior to the dorsal and anal fins. By 57-mm SL, these ridges are replaced by two or three median scutes, including those that extend onto the anterior margins of the caudal fin. One 57-mm-SL shortnose sturgeon also possessed two scutes along each side of the midline between the vent and anal fin with one median scute bordering the anterior margin of the anal fin. These paired vent-to-anal-fin scutes often fail to develop because they are often absent from many adults (Dadswell et al. 1984).

My fin ray and scute counts were not always consistent with those previously reported (Table 6). In particular, apparently complete pectoral fin ray counts observed for the larger larvae of both species were much lower than those previously reported for juveniles or adults. A difference of a few rays might be attributed to differing criteria if previous counts included the pectoral spine or individual rays incorporated in or associated with the spine. Despite the apparent completeness of pectoral fin ray counts by 30 mm SL, not all rays of the adult complement may be present yet, even for the largest larvae I examined. I observed considerable variation in pectoral fin ray counts for Atlantic sturgeon (27–38 irrespective of size between 30 and 136 mm SL) including differences of as many as 6 rays between right and left fins on some specimens. Counts of dorsal and caudal pterygiophores (fin ray supports) for Atlantic sturgeon were much higher than those reported by Ryder (1890) for juveniles or adults. Perhaps some fin supports fuse as the fish grow.

Size Relative to Developmental State

From the specimens studied, shortnose and Atlantic sturgeons are developmentally similar at hatching, but shortnose sturgeons are larger than Atlantic sturgeons, about 9–10 mm SL versus 7–9 mm SL. Shortnose sturgeons generally continue to be slightly larger at the same state of development than Atlantic sturgeons, at least through 60 mm SL (Table 7; Figures 2–7, 9).

The size difference at hatching was corroborated by recent hatchery experiments. For Atlantic sturgeons incubated at a mean temperature of 18°C, Smith et al. (1980) reported hatching in 5–6 d at a mean of 7.1 mm TL (1.9 mm SE). For shortnose sturgeons reared at 17°C, Buckley and Kynard (1981) reported hatching in 8 d at 9.5 mm TL. Dadswell et al. (1984) reported that Washburn and Gillis Associates (Fredericton, New Brunswick) reared shortnose sturgeons and observed hatching sizes of 7.3–11.3 mm, but that no specimens less than 8.0 mm survived. Taubert and Dadswell (1980) measured newly hatched (within 2 to 3 d) shortnose sturgeons collected from the Connecticut River at 8.0–12.5 mm TL.

Contrary to other observations of hatching size, Ryder (1890) reported that reared Atlantic sturgeons hatched at 11 mm TL. He illustrated a just-hatched specimen reported to measure 11.5 mm TL (Ryder's Figure 18). From total length and the concave shape of the dorsal fin fold, the illustrated specimen appears more typical of a recently hatched shortnose sturgeon. However, no other evidence in Ryder's report suggests that he might have reared shortnose sturgeons. For example, the diameter he reported for deposited unfertilized eggs, 2.6 mm, is consistent with observations for Atlantic sturgeons by Smith et al. (1980), 2–3 mm for unfertilized eggs, and by Jones et al. (1978), 2.0–2.9 mm for fertilized eggs. Shortnose sturgeon eggs are larger: Dadswell et al. (1984) and Buckley and Kynard (1981) reported diameters of 3.0–3.2 for ripe eggs and 3.5 mm for fertilized eggs, respectively. Yet, the hatching size Ryder reported is 2–5 mm larger than Smith et al. (1980) and I observed for Atlantic sturgeons, 1.5 mm larger than Buckley and Kynard (1981) reported for shortnose sturgeons, and at the upper end of the ranges Taubert and Dadswell (1980) and I observed for shortnose sturgeons. The hatching size reported by Ryder might have been an error. However, Ryder's emphasis on the size of his specimens relative to that of recently hatched *Acipenser ruthenus* suggests that he would have been particularly careful about this measurement. Vladykov and Greeley's (1963) remark that Atlantic sturgeons hatch at 11 mm was

TABLE 7.—Size (standard length, SL) and age at apparent onset of selected developmental events for shortnose sturgeon and Atlantic sturgeon, based on original observations at low-power magnification except as footnoted. Age from hatching is based on specimens reared at the Orangeburg National Fish Hatchery and might differ from similar observations on fish reared under other conditions. Observations on mesolarvae were limited and often size or age or both are reported only as greater than or less than that of available specimens. Rare or questionable values are enclosed in parentheses.

Developmental event	Shortnose sturgeon		Atlantic sturgeon	
	Size (mm SL)	Age (d)	Size (mm SL)	Age (d)
Hatching	(8[a])9–10(11)	0	(6)7–9(11[b])	0
Eyes pigmented	11–12(13[a])	1–3	10(11)	>1, 4
Pectoral bud formed	Prehatch	Prehatch	6, prehatch	0, prehatch
Pelvic bud formed	(11)12	3–4	10–11(12[c])	>1, 4
All yolk absorbed[d]	(13)14(15[a])	6	13(14[c])	6–7
All fin fold absorbed	~57	>30	>58, <67	~100
First element(s) present				
Fin ray supports				
Dorsal	11–(12)	1–3–(5)	10–11(12[c])	>1, 4
Anal	12–13	4–8	10–11	>1, 4
Caudal	>21, 24	>21, <25	20–21	>13, <29
Pectoral	(12)–14(15)	6–(8)	12–(13)	4–5
Pelvic	15–16	7–10	12	4–5
Fin rays				
Dorsal	>21, <24	>21, <25	>19[c], <29	>13, <29
Anal	34–37	>30	29–32	~29
Caudal	>24, <34	>30	>21, <29	>13, <29
Pectoral	>21, <24	>21, <25	(17)19[c], <29	(12)>13, <29
Pelvic	>24, <34	>30	>21, <29	>13, <29
Scute series				
Dorsal	>21, <24	>21, <25	>17, <20	>13, <29
Lateral	>21, <24	>21, <25	20–21	>13, <29
Ventral	>21, <24	>21, <25	>21, <29	>13, <29
Full complement present				
Fin ray supports				
Dorsal	(13)14–15	6–8	11–12	4
Anal	16–17	9	12–13	4–5
Caudal	>24, <34	>30	29–30	29
Pectoral	15	7–9	13–16	5–9
Pelvic	16–17	9–10	13–14	5–7
Fin rays				
Dorsal	>51, 57	>30	~29–30	–29
Anal	57	>30	>39, <47	>29, <100
Caudal	>57	>30	>116, <136	120
Pectoral	~36–37[e]	>30[e]	(30)~39–47[e]	>29, <100[e]
Pelvic	>41, <51	>30	>47, <58	>29, <100
Scute series				
Dorsal	>21, <24	>21, <25	20–21	>13, <29
Lateral	>41, <51	>30	>32, <39	~29
Ventral	>24, <34	>30	>21, <29	>13, <29

[a]Data modified or extended according to observations by Taubert and Dadswell (1980).
[b]Reported by Ryder (1890); inconsistent with other observations.
[c]Data modified or extended according to observations by Bath et al. (1981).
[d]Obvious yolk. Some yolk is probably retained in larger specimens but would require dissection to detect.
[e]Adult complement might not be acquired until larvae are larger and older.

not documented as to source but was probably based on Ryder (1890).

My observations on size at the onset of selected developmental events generally agree with those of Taubert and Dadswell (1980) and Bath et al. (1981). Among the differences incorporated in Table 7, Taubert and Dadswell (1980) reported that shortnose sturgeons acquire eye pigment and complete yolk absorption at larger sizes. Also, Bath et al. (1981) reported that Atlantic sturgeon larvae acquired pelvic fin buds and complete absorption of yolk at larger sizes and acquired first dorsal fin rays at smaller sizes than I observed.

Both Pekovitch (1979) and Taubert and Dadswell (1980) mistook median fin pterygiophores or basal structures (as in Figures 3–7) for incipient fin rays (see dorsal fins in Figure 7). The misrepresented fin ray counts were reiterated in a table of morphological and meristic parameters by Dadswell et al. (1984). Based on this misinterpretation of fin structure, Taubert and Dadswell (1980) also referred to 13–15-mm-TL shortnose sturgeons as mesolarvae, a developmental state not attained until the first median fin rays actually appear between 21 and 24 mm SL (Table 7).

In another instance of misapplication of developmental phase terminology, Bath et al. (1981) referred to 31.5–37-mm-TL Atlantic sturgeons as prejuveniles. However, the prejuvenile phase requires possession of the adult complement of fin rays in all fins (Mansueti and Hardy 1967). It is unlikely that the full complement of fin rays is present in any but the dorsal and possibly pectoral fins of specimens less than 40 mm TL (Table 7).

By 57 to 67 mm, both species have met all criteria for transition to the juvenile period (Snyder 1976, 1983) except attainment of the adult complement of caudal fin rays (Table 7). A full complement of caudal fin rays does not occur on Atlantic sturgeons until nearly twice that size (greater than 116 mm SL) but might be achieved by shortnose sturgeons at a much smaller size. Both species remain mesolarvae until the juvenile period; by definition, they do not pass through the metalarval phase typical of teleosts.

Pigmentation

Although pigmentation is similar for recently hatched protolarvae of both species (Figures 2, 9), it is the most obvious diagnostic character for confidently segregating all shortnose and Atlantic sturgeon larvae more than (and sometimes equal to) 12 mm SL (Figures 3–8). The ventral surface of the abdomen of shortnose sturgeon larvae, at and below the level of the pectoral fins or fin buds, is white (practically no melanophore pigmentation). This includes the ventral surface of the pectoral fin and later the ventral series of scutes. On Atlantic sturgeon larvae with partially assimilated yolk, melanophores cover the ventrolateral surfaces of the abdomen and the base of the pectoral fin. After the obvious yolk supply is exhausted, melanophores rapidly spread over most of the ventral surface of the abdomen, pectoral fins, and gill covers. As Atlantic sturgeons grow beyond 30 to 40 mm, melanophores on the ventral abdomen become more widely spaced. The surface becomes quite pale but remains at least sparsely pigmented with a uniform speckling of melanophores, even on specimens as large as 136 mm SL. The appearance of pigmentation (density and degree of melanophore expansion) can vary considerably on both species (compare Figure 5, bottom, with Figure 8), but the specific generalized pattern of distribution is consistent at least through the larval period. Scott and Crossman (1973) and Vladykov and Greeley (1963) described adults of both species as white on the ventral surface of the body.

As a diagnostic character for identifying protolarvae without yolk and mesolarvae, abdominal pigmentation was overlooked or unspecified in previous descriptions. Pekovitch (1979) noted that for 16–18-mm-TL shortnose sturgeon, "pigmentation was absent ventrally from the mouth to the anus. . . ." However, his three-view drawings of 16- and 32-mm shortnose sturgeons do not clearly illustrate the white ventral surface and might be misleading because pigmentation cannot be readily distinguished from dot-pattern shading that covers all surfaces. Taubert and Dadswell (1980) also noted that shortnose sturgeon larvae, 13–15 mm TL, had a "light ventrum", but Bath et al. (1981) made no mention of ventral pigmentation. Still, lateral-view photographs of both species in Bath et al. (1981) and of shortnose sturgeons in Taubert and Dadswell (1980) illustrate the difference in ventrolateral pigmentation. A lateral-view photograph in Buckley and Kynard (1981) also illustrates the distinctly white ventrolateral surface of shortnose sturgeon mesolarvae. Finally, Dovel and Berggren (1983) noted that there are differences in pigmentation but did not elaborate on the nature of these differences and suggested the character could be confusing. However, the ventral-view photographs of both species in Dovel (1979) and Dovel and Berggren (1983) clearly illustrate the differences.

Correlation with Capture Date and Location

In at least some cases, identity of recently hatched sturgeon larvae can be corroborated with information on date and location of capture. Depending on the specific river system, shortnose sturgeons usually begin spawning about a month earlier, have a shorter spawning period, spawn at cooler water temperatures (9–15°C versus 13–18°C), and, in some rivers, use more upstream locations for spawning than Atlantic sturgeons (Borodin 1925; Vladykov and Greeley 1963; Scott and Crossman 1973; Jones et al. 1978; Dadswell 1979; Taubert 1980; Bath et al. 1981; Dovel and Berggren 1983; Dadswell et al. 1984).

The value of capture date and location for

identification of larvae in the Hudson River was documented by Bath et al. (1981). Using primarily mouth width characters, they identified shortnose sturgeon protolarvae without yolk only in collections from an upstream, freshwater portion of the river at river kilometer 235 (near Albany) during May and similar-size Atlantic sturgeons only in downstream collections from oligohaline waters from km 64 to 106 (Stony Point to Chelsa) during June and July. Based on these results and other information on species distributions, movements, and probable spawning seasons and grounds in the Hudson River, they tentatively identified yolk-bearing larvae collected in the oligohaline portion of the river from km 60 to 126 (Haverstraw to Hyde Park) during June and July as Atlantic sturgeons.

Recently hatched protolarvae from the Savannah River also were segregated by collection date (Figure 9). All those suspected to be shortnose sturgeons, based on larger size relative to state of development and, in some cases, dorsal finfold depth and shape, were collected in March of 1982 and 1983 at river temperatures usually between 11 and 13°C. Those suspected to be Atlantic sturgeons were taken, with one exception, in April and May at river temperatures usually between 15 and 22°C. The single exception was a specimen captured in August at 21°C. Based on observations of running-ripe fish in the Savannah and other rivers in South Carolina, there appeared to be a second population of Atlantic sturgeon that spawns in August to October (Smith, personal communication). Unlike sturgeon larvae in the Hudson River, recently hatched larvae of both species were collected in the same portions of the Savannah River, mostly adjacent to the Savannah River Plant (km 242 to 253), but some were taken as far downstream as Porter's Landing (km 113) (Paller et al. 1984; note that in their Table 2–22, the larva collected on March 29, 1983, should be listed as shortnose sturgeons, not Atlantic sturgeons).

Conclusions

Although larval development is similar for both species, shortnose sturgeon larvae at a given developmental state are usually larger than Atlantic sturgeon larvae, at least through 60 mm SL. Both species absorb most of their yolk by 13 or 14 mm SL and remain protolarvae until the first median fin rays appear in the dorsal fin between 19 and 24 mm SL. By 60–70 mm SL, all fin fold is lost and, if one assumes that ray formation in the pectoral fins is complete, the adult complement of fin rays is present in all but the caudal fin. As a result, the metalarval phase typical of teleosts is eclipsed. Transition from the mesolarval phase to the juvenile period occurs when the adult complement of caudal rays is acquired, probably between 116 and 136 mm SL for Atlantic sturgeons and possibly at a smaller size for shortnose sturgeons.

Morphological criteria for separation of recently hatched sturgeon larvae consist of size relative to state of development and depth and shape of the dorsal fin fold. However, these criteria might not be conclusive, and identities based on them should be considered tentative. Shortnose sturgeons hatch at a larger size than Atlantic sturgeons, usually at 9–10 mm SL (full range, 8–11 mm SL) versus 7–9 mm SL (full range, 6–9 mm SL exclusive of one questionable report of 11 mm TL). The anterior portion of the dorsal fin fold of yolk-bearing shortnose sturgeon protolarvae is often slightly concave in shape and shallower than the typically convex fin fold of Atlantic sturgeons. Specimens measuring 8–12 mm SL with a dorsal fin fold about a third or less of body depth (excluding fin folds) at the future origin of the pelvic fin buds (about two-thirds the distance from yolk sac to vent) are shortnose sturgeons. Specimens with dorsal finfold depths greater than a third of body depth could be either species. Many sturgeon larvae approaching yolk depletion can be identified on the basis of ventrolateral pigmentation, inter–lip-lobe distance, or both.

Suspected identity of recently hatched larvae should be corroborated with information on the date and, in some cases, location of capture. Shortnose sturgeons typically begin spawning about a month earlier and, in some rivers, much farther upstream than Atlantic sturgeons.

After the yolk is consumed, Atlantic sturgeon larvae are easily distinguished from shortnose sturgeons (barring possible hybrids) by the presence of melanophore pigmentation on the ventrolateral to ventral surfaces of the abdomen and an inter–lip-lobe distance less than 20% of mouth width (lips included). Shortnose sturgeon larvae are white (with almost no melanophores) on the ventrolateral and ventral surfaces of the abdomen and have an inter–lip-lobe distance greater than 25% of mouth width.

Although not as definitive as the above characters, mouth width (including lips) is diagnostic for specimens over 14 mm SL. For protolarvae, mouth width as a percentage of head width, measured in line with the mouth, is greater than 65% (66–88%) for shortnose sturgeons and less than 66% (57–65%) for Atlantic sturgeons. For mesolarvae, the same criteria are greater than 60% (62–74%) and less than 60% (56–59%), respectively.

The number of pelvic fin rays and, to a lesser extent, number of anal fin rays are also diagnostic for larvae over 50–60 mm SL. Pelvic fin ray counts are 17–22 for shortnose sturgeons and 26–33 for Atlantic sturgeons. Anal ray counts are 18–24 and 23–30, respectively.

The most diagnostic morphometric and meristic characters for mesolarvae of shortnose sturgeons and Atlantic sturgeons should be equally definitive and obvious for distinguishing juvenile and adult specimens. In particular, inter–lip-lobe distance relative to mouth width (specific percentages might differ) and pelvic ray counts should be considered as supplements or alternatives to primary characters used in recent keys (e.g., Vladykov and Greeley 1963; Moore 1968; Scott and Crossman 1973; Dadswell et al. 1984).

Acknowledgments

I thank E. K. Dingley and R. Lindsey of the U.S. Fish and Wildlife Service; T. I. J. Smith and D. E. Marchette of the South Carolina Wildlife and Marine Resources Department; D. V. Osteen of Environmental and Chemical Sciences, Incorporated, Aiken, South Carolina; M. J. Dadswell of Acadia University, Wolfville, Nova Scotia; D. W. Bath of the Laboratory of Environmental Studies of the New York Medical Center; and D. M. Carlson of the New York Department of Environmental Conservation for the loan of specimens. Bath and Carlson also provided valuable references and supplemental data. C. L. Bjork prepared the illustrations. E. Weddel and B. J. Aldrich typed the manuscript. C. A. Carlson, R. T. Muth, and T. I. J. Smith critically reviewed the manuscript. This paper is based on a final report of the same title prepared for E. I. du Pont de Nemours and Company, Savannah River Laboratory, Aiken, South Carolina (July 20, 1984). R. Matthews and J. B. Gladden of the Savannah River Laboratory arranged for research and publication support. The study was funded by the U.S. Department of Energy under contract DE-AC09-76SR00001.

References

Bath, D. W., J. M. O'Connor, J. B. Alber, and L. G. Arvidson. 1981. Development and identification of larval Atlantic sturgeon (*Acipenser oxyrhynchus*) and shortnose sturgeon (*A. brevirostrum*) from the Hudson River estuary, New York. Copeia 1981: 711–717.

Bigelow, H. B., and W. C. Schroeder. 1953. Fishes of the Gulf of Maine. U.S. Fish and Wildlife Service Fishery Bulletin 53(74).

Borodin, N. 1925. Biological observations on the Atlantic sturgeon (*Acipenser sturio*). Transactions of the American Fisheries Society 55:184–190.

Buckley, J., and B. Kynard. 1981. Spawning and rearing of shortnose sturgeon from the Connecticut River. Progressive Fish-Culturist 43:74–76.

Dadswell, M. J. 1979. Biology and population characteristics of the shortnose sturgeon, *Acipenser brevirostrum* LeSueur 1818 (*Osteichthyes: Acipenseridae*), in the Saint John River estuary, New Brunswick, Canada. Canadian Journal of Zoology 57: 2186–2210.

Dadswell, M. J., B. D. Taubert, T. S. Squiers, D. Marchette, and J. Buckley. 1984. Synopsis of biological data on shortnose sturgeon, *Acipenser brevirostrum* LeSueur 1818. NOAA (National Oceanic and Atmospheric Administration) Technical Report NMFS (National Marine Fisheries Service) 14. [Also FAO (Food and Agriculture Organization of the United Nations) Fisheries Synopsis 140.]

Dovel, W. L. 1979. The biology and management of shortnose and Atlantic sturgeon of the Hudson River. New York Department of Environmental Conservation, Report AFS9-R, Albany.

Dovel, W. L., and T. J. Berggren. 1983. Atlantic sturgeon of the Hudson estuary, New York. New York Fish and Game Journal 30:140–173.

Fuiman, L. A. 1982. Correspondence of myomeres and vertebrae and their natural variability during the first year of life in yellow perch. Pages 56–59 *in* C. F. Bryan, J. V. Conner, and F. M. Truesdale, editors. The fifth annual larval fish conference. Louisiana Cooperative Fishery Research Unit and the School of Forestry and Wildlife Management, Louisiana State University, Baton Rouge.

Hildebrand, S. F., and W. C. Schroeder. 1928. Fishes of the Chesapeake Bay. U.S. Bureau of Fisheries Bulletin 43 (part 1).

Johnson, J. E. 1987. Protected fishes of the United States and Canada. American Fisheries Society, Bethesda, Maryland.

Jones, P. W., F. D. Martin, and J. D. Hardy, Jr. 1978. Development of fishes of the mid-Atlantic Bight, volume 1. U.S. Fish and Wildlife Service Biological Services Program FWS/OBS-78/12.

Jordan, D. S., and B. W. Evermann. 1896. The fishes of North and Middle America, U.S. National Museum Bulletin 47 (part 1).

Lee, D. S., C. R. Gilbert, C. H. Hocutt, R. E. Jenkins, D. E. McAllister, and J. R. Stauffer, Jr. 1980. Atlas of North American freshwater fishes. North Carolina State Museum of Natural History, Raleigh.

Lippson, A. J., and R. L. Moran. 1974. Manual for identification of early developmental stages of fishes of the Potomac River estuary. Report to

Maryland Power Plant Siting Program, PPSP—MP-13, Annapolis.

Mansueti, A. J., and J. D. Hardy, Jr. 1967. Development of fishes of the Chesapeake Bay region: an atlas of egg, larval, and juvenile stages, part 1. Natural Resources Institute, University of Maryland, Baltimore.

Moore, G. A. 1968. Fishes. Pages 21–165 *in* W. F. Blair, A. P. Blair, P. Brodkorb, F. R. Cagle, and G. A. Moore. Vertebrates of the United States, 2nd edition. McGraw Hill, New York.

Paller, M., J. O'Hara, V. Osteen, W. Specht, and H. Kania. 1984. Annual report on the Savannah River Aquatic Ecology Program, September 1982–August 1983, volume 1. Report to E. I. du Pont de Nemours and Company, Savannah River Laboratory, Aiken, South Carolina.

Pekovitch, A. W. 1979. Distribution and some life history aspects of the shortnose sturgeon (*Acipenser brevirostrum*) in the upper Hudson River estuary. Report to New York State Electric and Gas Corporation, Binghamton, New York.

Ryder, J. A. 1890. The sturgeons and sturgeon industries of the eastern coast of the United States, with an account of experiments bearing upon sturgeon culture. U.S. Fish Commission Bulletin 8(1888): 231–328.

Scott, W. B., and E. J. Crossman. 1973. Freshwater fishes of Canada. Fisheries Research Board of Canada Bulletin 184.

Seifert, R. E. 1969. Characteristics for separation of white and black crappie larvae. Transactions of the American Fisheries Society 98:326–328.

Smith, T. I. J. 1985. The fishery, biology, and management of Atlantic sturgeon, *Acipenser oxyrhynchus*, in North America. Environmental Biology of Fishes 14:61–72.

Smith, T. I. J., E. K. Dingley, and D. E. Marchette. 1980. Induced spawning and culture of Atlantic sturgeon. Progressive Fish-Culturist 42:147–150.

Snyder, D. E. 1976. Terminologies for intervals of larval fish development. U.S. Fish and Wildlife Service Biological Services Program FWS/OBS-76/23:41–58.

Snyder, D. E. 1981. Contribution to a guide to the cypriniform fish larvae of the upper Colorado River system in Colorado. U.S. Bureau of Land Management, Biological Sciences Series 3, Denver, Colorado.

Snyder, D. E. 1983. Fish eggs and larvae. Pages 165–197 *in* L.A. Nielsen and D. L. Johnson, editors. Fisheries techniques. American Fisheries Society, Bethesda, Maryland.

Taubert, B. D. 1980. Reproduction of shortnose sturgeon (*Acipenser brevirostrum*) in Holyoke Pool, Connecticut River, Massachusetts. Copeia 1980: 114–117.

Taubert, B. D., and M. J. Dadswell. 1980. Description of some larval shortnose sturgeon (*Acipenser brevirostrum*) from the Holyoke Pool, Connecticut River, Massachusetts, U.S.A., and the Saint John River, New Brunswick, Canada. Canadian Journal of Zoology 58:1125–1128.

Vladykov, V. D., and J. R. Greeley. 1963. Order Acipenseroidei. Pages 24–60 *in* Fishes of the western North Atlantic. Part 3. Sears Foundation for Marine Research, Yale University, New Haven, Connecticut.

American Fisheries Society Symposium 5:31–48, 1988

Spawning of Atlantic Herring in the Gulf of St. Lawrence

S. Messieh[1]

Science Branch, Department of Fisheries and Oceans Gulf Region, Post Office Box 5030
Moncton, New Brunswick E1C 9B6, Canada

Abstract.—Spawning locations of Atlantic herring *Clupea harengus harengus* in the Gulf of St. Lawrence were identified by aerial photographic surveys of gill-net distributions. The spawning locations have been relatively consistent over time. Spawning beds were surveyed by scuba divers in the springs of 1980–1984 and in the autumns of 1985 and 1986. The spawning time and duration inferred from the maturity stages showed a definite regularity. Spring spawners arrived on the spawning grounds in late April or early May, and autumn spawners arrived in August. A significant inverse relationship existed between the time of arrival of Atlantic herring on the spawning grounds during spring and temperature. The spring-spawning beds were shallow (0.8–4.7 m). The substrate consisted of bedrock and rubble; Irish moss *Chondrus crispus* and rockweeds *Fucus* sp. were the dominant macrophytes. Mean bottom temperature ranged from 6.0 to 7.5°C during the incubation period, which lasted for 17–20 d. Density of egg deposition varied from 1,200 to 95,800 eggs/m^2, and patches of eggs did not exceed five layers. The main predator during this period was the winter flounder *Pseudopleuronectes americanus*, in which the highest average number of eggs per stomach was 6,480. The autumn-spawning beds were 10–25 m deep. The substrate consisted of bedrock with cobble and negligible vegetation. In one area in 1985, autumn eggs were deposited on bare rock in a thick sheet of up to 30 layers, and mass mortality of the lower eggs occurred. Densities of egg deposition were as high as 7.9×10^6 eggs/m^2 in 1985 and 3.8×10^6 eggs/m^2 in 1986. Mean bottom temperature ranged from 14.8 to 16.0°C during the incubation period, which lasted 6–8 d. The main predators during this period were winter flounder and (reported for the first time) Atlantic mackerel *Scomber scombrus*. Atlantic herring spawning behavior was studied in autumn 1986 with underwater video equipment.

Annual surveys of spawning bed of Pacific herring *Clupea harengus pallasi* on the coast of British Columbia have been conducted since 1937 (Humphreys and Hourston 1978). Since the 1950s, scuba and aerial photography of spawning areas have increased the accuracy of these surveys (Haegele and Schweigert 1985). In the northwestern Atlantic Ocean, one semiquantitative scuba diving survey was made of a spawning bed of Atlantic herring *Clupea harengus harengus* in 1962 (Tibbo et al. 1963), but there were no others until 1980.

The Gulf of St. Lawrence is the only area in the northwestern Atlantic where both spring- and autumn-spawning Atlantic herring support major commercial fisheries. In 1980, the Department of Fisheries and Oceans, Canada, initiated herring spawning bed projects in the southern Gulf of St. Lawrence (Figure 1) to provide biological data for stock assessments (Messieh et al. 1985). Scuba-diving surveys were carried out in the springs of 1980, 1981, 1983, and 1984 and in the autumns of 1985 and 1986. The spawning-bed surveys allowed in situ observations of the spawning conditions of the two populations in this area.

In the present paper, I present survey data on the spawning locations and times, character of the spawning beds, intensity of egg deposition, and factors that regulate the arrival of Atlantic herring on the spawning grounds.

Methods

Aerial surveys of the spawning grounds.—The spawning locations of the Atlantic herring populations were identified by aerial photography of the gill nets on the spawning grounds during the fishing season (Messieh and Pottle 1986). Two aerial cameras were used on different occasions: a K17B with 15-cm focal length and a yellow filter (minus blue) and a Wild RCB with a 15-cm focal length and a yellow, antivignetting filter. Intensity of fishing gear was expressed as number of standard gill nets/km^2. Spawning locations of the autumn populations were collected from questionnaires and interviews with fishery officers and experienced fishermen, because it was difficult to use aerial surveys for the autumn fishery.

[1]Present address: Department of Fisheries and Oceans, Bedford Institute of Oceanography, Habitat Ecology Division, Post Office Box 1006, Dartmouth, Nova Scotia B2Y 4A2, Canada.

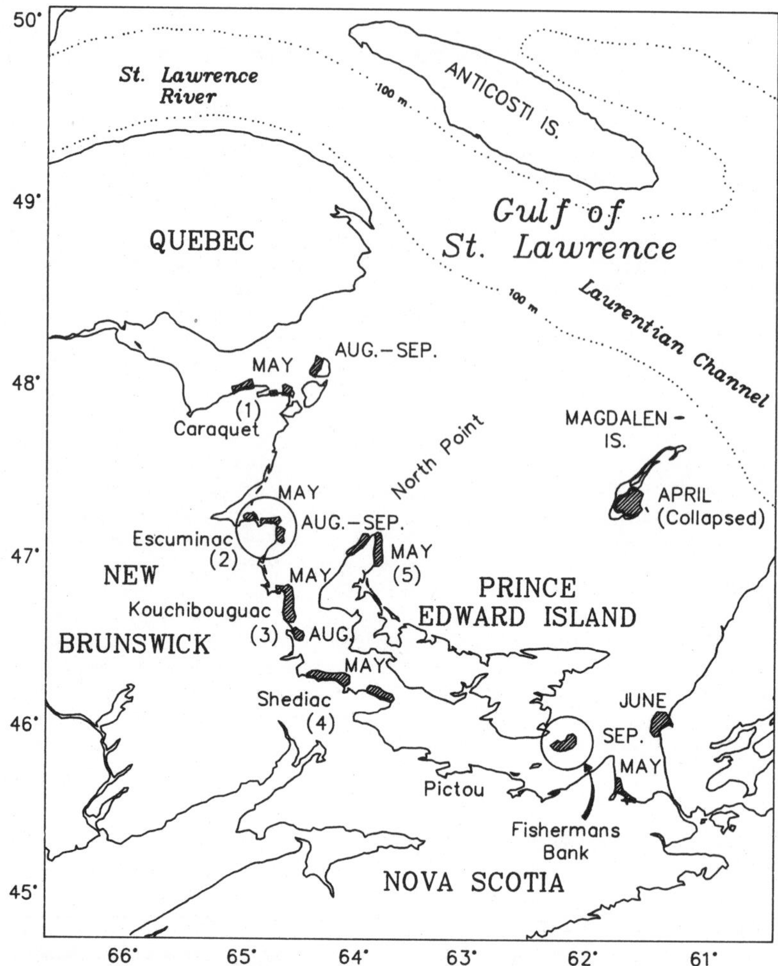

FIGURE 1.—Spring- and autumn-spawning grounds of Atlantic herring in the Gulf of St. Lawrence. Circled spawning locations were investigated by scuba divers.

Duration of the spawning period.—The durations of the spawning periods for spring-spawning and autumn-spawning populations were determined from analysis of maturity stages in many samples of Atlantic herring, comprising more than 30,000 fish, which were collected during the prespawning and spawning periods in 1966–1984. Analysis of weekly samples from various fishing areas enabled the identification of spawning peaks with a good degree of precision. Spawning was considered to have reached its peak when more than 50% of the Atlantic herring in the samples were in the ripe and running condition. The relationship between the arrival time of Atlantic herring on the spawning grounds near Magdalen Islands and mean surface temperatures available for a 15-year period was examined by regression analysis.

Scuba surveys of the spawning beds.—Scuba-diving surveys of spawning beds were carried out in the springs of 1980, 1981, 1983, and 1984 and in the autumns of 1985 and 1986. The spring surveys were carried out on a major spawning ground at Escuminac in Miramichi Bay, New Brunswick. The autumn spawning surveys were conducted on Fishermans Bank off southeastern Prince Edward Island (Figure 1). These locations were chosen because they are the centers of major herring fisheries in the southern Gulf of St. Lawrence.

A chartered fishing vessel (14 m long) was used in most of the surveys. Location of the sample stations was determined by a LORAN C navigation system, which has a precision of ±15 m. Surface and bottom temperatures were recorded and occasionally, continuous temperature recording by Ryan thermographs and salinity observations were taken.

Each survey was carried out by three or four divers. The search began with bounce dives at regular intervals on the potential spawning area. Each dive was of 3–5 min duration and consisted of a quick descent to the bottom along the anchor cable and a search of the substrate within a radius of 4–5 m. Data on depth, substrate type, and algal species present were recorded at each station. When spawn was identified, a search grid with 100-m squares was plotted on a hydrographic chart of the study area, and LORAN C bearings were determined for each station. Divers located and mapped the perimeter of the spawning bed. In 1986, an underwater video camera system (EDO Western Camera with a low-light tube and Cosmicar 10-mm wide-angle lens and monitor) was used to assist the search operation. Contour maps of egg distributions were generated by computer with a GPCPII plotting program.

At each station within the identified spawning bed, divers collected Atlantic herring eggs and algae within a randomly selected quadrat (0.25 m^2). All loose material scraped from the substrate was collected with a diver-operated airlift sampler. In case of heavy deposition, the airlift was not used and the samples were physically cut out of the egg sheet within the quadrat. Samples were placed in buffered 5% formalin for laboratory examination.

Egg counts were estimated from wet samples in the case of spring spawning because of the small number of eggs involved. Egg number was determined either by direct count or, with large samples, by extrapolation from subsamples created by a plankton splitter.

In the case of autumn spawning, a large number of eggs in the shape of a thick sheet were involved, and counts were made from dried samples. Eggs were thoroughly cleaned and freed from gravel and then placed in an oven (60°C) to dry to constant weight in about 16 h. Subsamples of about 1,000 eggs were weighed with a high-precision Mettler balance and accurately counted; then estimates of total number of eggs per quadrat were made. Two to three counts were usually taken for validation, and differences in egg counts did not exceed 1%. Egg density was expressed in number of eggs/m^2.

Developmental stages of Atlantic herring eggs.—Samples of Atlantic herring eggs were collected for examination of developmental stages. The eggs were classified according to the scale of Baxter (1971) as follows.

Stage 1: from fertilization to the formation of the blastula.

Stage 2: gastrula stage; head mesoderm not yet differentiated from the trunk mesoderm.

Stage 3: head region shows some organization; head and trunk still attached to yolk.

Stage 4: keel of anterior portion of head rises out clear of yolk; tail has grown out free; eyes may show light pigmentation.

Stage 5: prehatching stage; embryo completely formed; eyes heavily pigmented.

Fish predators.—Counts of benthic fish predators, primarily winter flounder *Pseudopleuronectes americanus*, within visual range of the diver were made along 200–300-m transect lines randomly positioned on the spawning bed. Samples of fish were collected by hand spear and saved for stomach analysis. In the autumn spawning surveys, schools of Atlantic mackerel *Scomber scombrus* were found on the spawning bed and sampled by hand line. Stomach contents were estimated volumetrically after eggs had been separated from other gut contents.

Results

Spawning Grounds

From aerial photographs taken in May of 1980, 1981, and 1983, the major spring Atlantic herring fisheries in the southern Gulf of St. Lawrence were located in the general vicinity of Caraquet, Escuminac, Kouchibouguac, and Shediac in New Brunswick and at North Point in Prince Edward Island (Figure 2). All these fisheries have been important historically. The gill-net distributions during this time were essentially the same. In general, the spawning grounds (as inferred from the gill-net locations) have remained approximately the same for many years except for small variations in some areas. For example, in the Caraquet area, the fishery has shifted a few kilometers westward from the vicinity of Caraquet to Belloni Point; in the Shediac area, gill-net distributions were variable, but the concentration was near Foxes Point. The Escuminac spawning grounds have remained stable through time and these Atlantic herring concentrations have supported the largest inshore Atlantic herring fishery in the southern Gulf of St. Lawrence. There was little variation in the gill-net distributions over the 1980–1983 period in the Kouchibouguac and North Point areas.

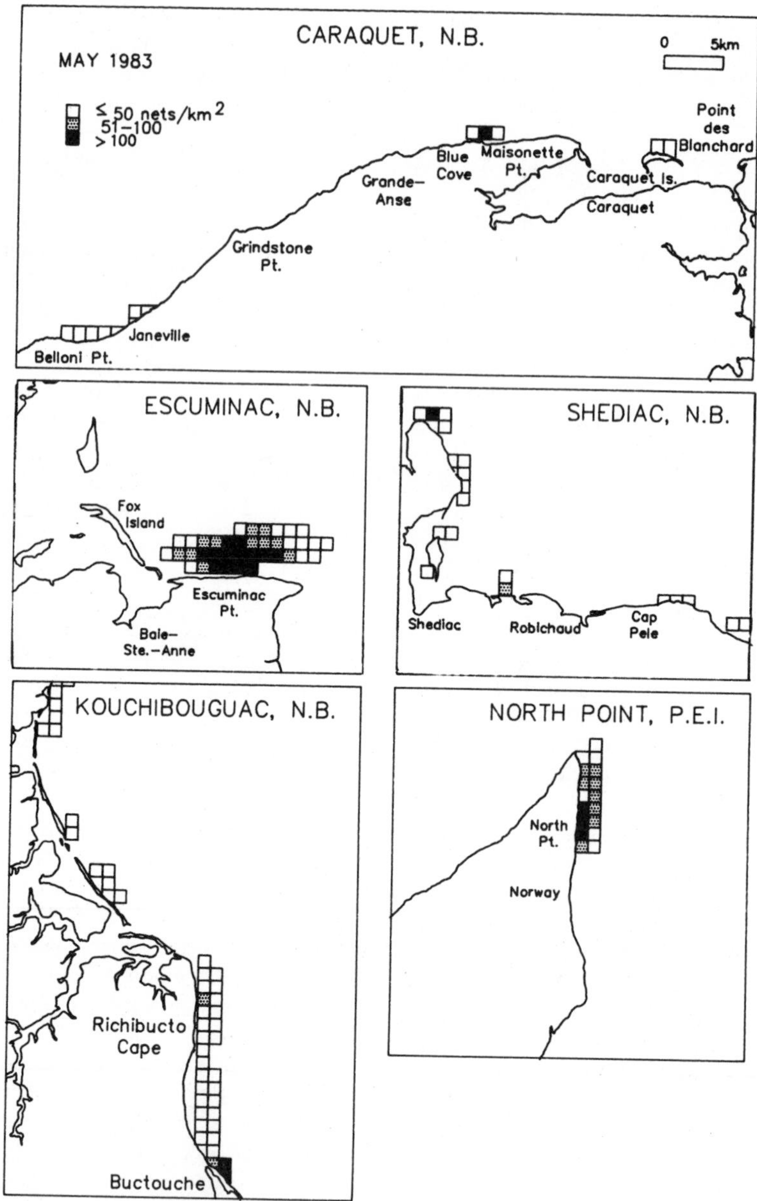

FIGURE 2.—Spring-spawning locations of Atlantic herring in the southern Gulf of St. Lawrence as inferred from aerial photographic surveys of herring gill-net distributions. N.B. is New Brunswick; P.E.I. is Prince Edward Island.

Spawning Duration

The spawning periods for the spring and autumn spawners and the weeks in which peak spawning occurred (Figure 3) were estimated from the Atlantic herring samples for the five major spawning areas of the southern Gulf of St. Lawrence during 1966–1984. The Magdalen Islands fishery is a spring-spawner fishery. In the 1960s, spawning occurred from mid-April to late May with peaks from the last week of April to the second week of May. In the 1970s, there was no evidence of peaks in spawning. The lack of major spawn coincided with the collapse of the fishery and the dominance of nonspawning fish in the

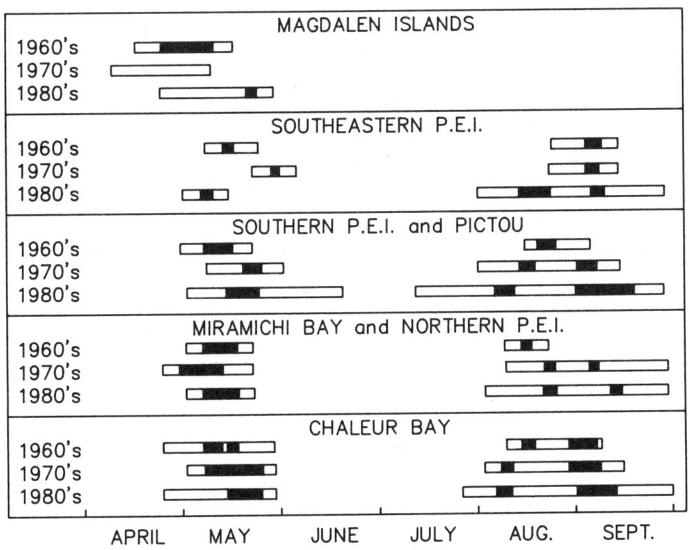

FIGURE 3.—Time and duration of spring and autumn spawning by Atlantic herring in five areas of the Gulf of St. Lawrence by decades for 1966–1984 as inferred from analysis of maturity stages. Black bars indicate peak spawning, when more than 50% of the fish were mature. P.E.I. is Prince Edward Island.

limited catches of that period. The population has not recovered yet, but some samples of mature Atlantic herring taken in 1984 showed peak spawning in late May.

Along southeastern Prince Edward Island, the spring-spawning period generally extended over a period of 3 weeks with considerable variation in the timing of the peak spawning, which ranged from the second week of May in the 1980s to the first week of June in the 1970s (Figure 3). Autumn (or late-summer) spawning in this area extended over a much longer period in the 1980s (9 weeks) than in the 1970s (4 weeks); major spawning occurred in mid-August and early September during the 1980s, but only in early September during the 1960s and 1970s.

In the southern Prince Edward Island–Pictou area (Figure 3), spring spawning usually occurred during May in the 1960s and 1970s, but there was an extension of the spawning period to late June in the early 1980s, although the peak period has remained in May through time. The autumn-spawning period in this area was rather short (4 weeks) in the 1960s, became longer in the 1970s (7 weeks), and was still longer (11 weeks), from mid-July to late September, in the 1980s. Spawning peaks occurred in early August and in early to mid-September in the 1970s and 1980s.

The Escuminac grounds in Miramichi Bay have seen very little variation in the timing and duration of spring spawning (Figure 3). In some years, the spawning period was limited to only 2 weeks. Autumn spawning occurred off northwestern Prince Edward Island and extended from 3 weeks (peak in mid-August) in the 1960s to 8–9 weeks in August and September of the 1970s and 1980s, when peaks occurred in late August and early to mid-September.

In the Chaleur Bay area (Figure 3), spring spawning extended from late April to late May in all three decades with considerable regularity. Autumn spawning period in this area increased in duration from about 4 weeks in the 1960s to 10 weeks in the 1980s, exhibiting bimodal spawning peaks in early August and early September.

Spring Spawning Beds

The 4 years of diving surveys showed that the locations of spring spawning beds in Miramichi Bay changed little from year to year (Table 1). In 1980, the spawning bed was about 3.5 km off Escuminac wharf; in 1981–1984, it lay within a 1-km radius of the wharf (Figure 4).

In 1980, a single spawning bed was located on May 17 near Fox Island, 3.5 km northwest of Escuminac wharf. Mean (offshore) width of the spawning bed was 250 m, and the bed extended for about 450 m along the shoreline and resulted in an estimated area of 87,500 m^2. Eggs were distributed at water depths of 0.9–4.3 m (mean low

TABLE 1.—Location, depth, and area of Atlantic herring spawning beds in the Gulf of St. Lawrence surveyed in 1980–1986.

Year	Starting date of survey	Date of spawn discovery	Spawning bed	Location	Depth range (m at mean low water)	Spawning area (10^3 m^2)
1980	May 12	May 17	1	Miramichi Bay	0.9–4.3	88
1981	Apr 25	May 3	1	Miramichi Bay	2.0–4.7	30
		May 15	2		2.0–4.7	13
1983	Apr 27	May 20	1	Miramichi Bay	0.8–3.3	880
1984	May 10	May 12	1	Miramichi Bay	1.4–4.3	560
		May 14	2		2.6–4.4	445
		May 25	3		1.2–2.9	1,100
1985	Aug 30	Sep 6	1	Fishermans Bank	10–15	286
		Sep 9	2		10–15	489
		Sep 13	3		17–23	239
		Sep 18	4		20–25	247
		Sep 18	5		20–25	560
1986	Aug 27	Sep 4	1	Fishermans Bank	15–20	1,100

water, MLW); areas of high concentration were 1.4–4.0 m deep (Figure 5). All egg deposition was on algal cover except in the vicinity of gill nets, where a few eggs were found loose on bare substrate.

In 1981, the divers located a bed on May 3 about 200 m northwest of the Escuminac wharf and about 430 m offshore. Thirty-five quadrat samples were collected from May 9 to 14. Most stations were sampled twice; however, the dangerous situation created by poor visibility, strong currents, and a fixed gill-net set through the center of the site prevented repeated sampling of stations close to the net. The spawning bed was irregular in shape with a width of about 100 m and a length of about 125 m (Figure 5). The area of the spawning bed was estimated at 12,000 m^2. Another area of egg deposition estimated at 30,000 m^2 was located on May 15 about 200 m inshore of the first bed. Most of egg deposition was in the form of a few scattered eggs on algal cover. The beds were in a water depth ranging from 2.0 to 4.7 m (MLW).

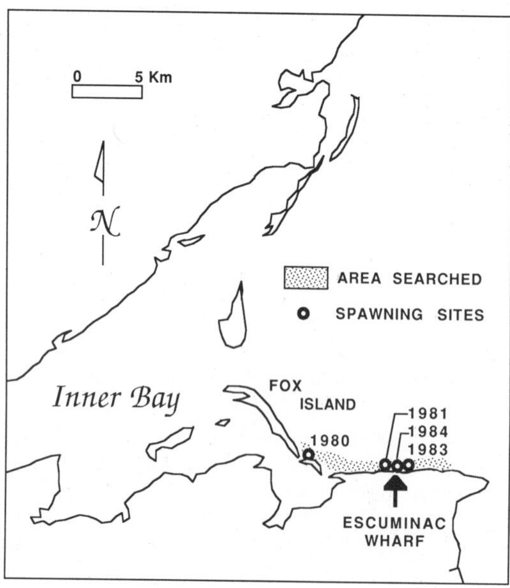

FIGURE 4.—Map of Miramichi Bay, New Brunswick, showing locations of Atlantic herring spawning beds, 1980–1984.

In 1983, a single bed was found on May 20 0.8 km east of Escuminac wharf. The bed extended 2,200 m along the shoreline with a maximum width of about 650 m (mean width of 400 m) and covered an area of about 0.88 km^2 (Figure 5). Eggs were distributed over a water depth range of 0.8–3.3 m. The perimeter of the bed was correlated with the extent of algal cover. Egg density varied, ranging from a few scattered eggs to small clumps 4–5 layers deep.

In 1984, three large spawning beds were found near the Escuminac wharf, not far from the spawning bed found in 1983 (Figure 5). The first bed was discovered on May 12, about 100 m west of the wharf at depths of 1.4–4.3 m (MLW). The second bed was found on May 14 400 m east of the wharf, in an area that had been barren the day before based on the diving log. Eggs at this site had been deposited primarily on bare rock substrate at depths of 2.6–4.4 m (MLW). A third spawning bed was found further east on May 25. This was approximately the same distance from

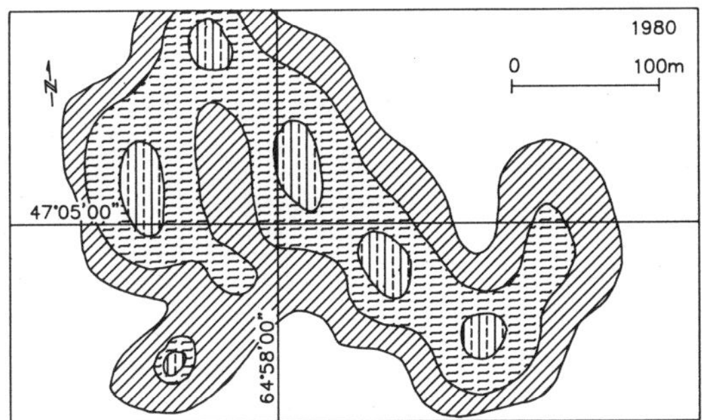

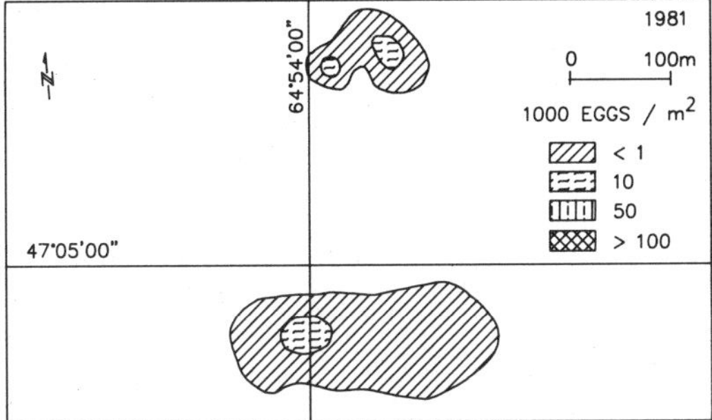

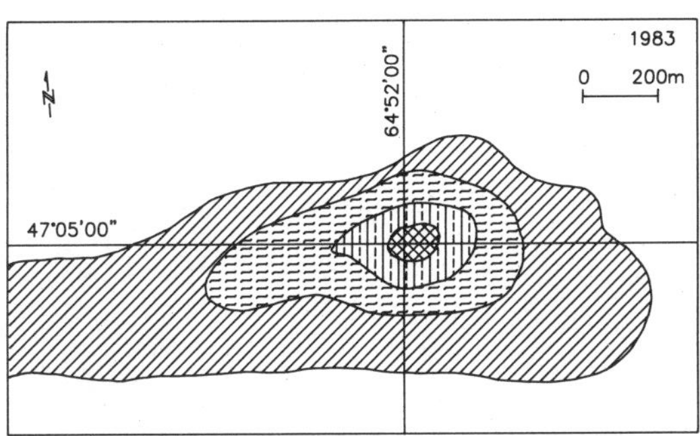

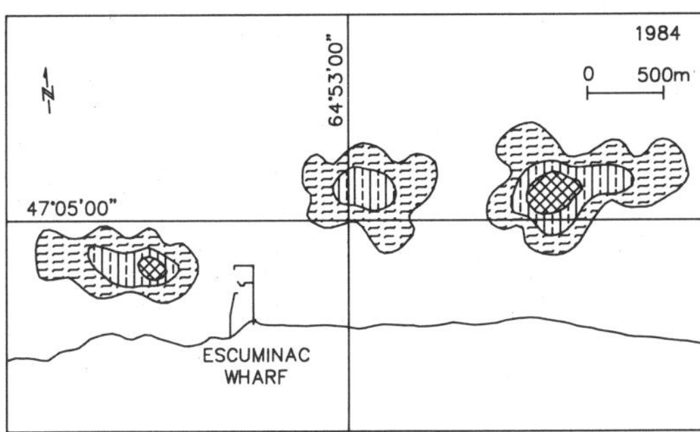

FIGURE 5.—Contour plots of Atlantic herring egg density of the spawning beds of Escuminac, Miramichi Bay, 1980–1984.

the wharf as the bed surveyed in 1983 but further from shore. Eggs at this site were at water depths of 1.2–2.9 m MLW. The developmental stage of eggs of this site indicated that spawning had occurred during the same time period as at the two other beds. The estimated areas of the three spawning beds were 560,000, 445,000 and 1,100,000 m², respectively (Table 1).

TABLE 2.—Percent macrophyte species (wet weight) on Atlantic herring spawning beds in spring 1984, Miramichi Bay.

Macrophyte	Spawning bed 1	2	3
Epiphytes	0.2	0	0.3
Chondrus sp.	96.2	96.3	93.1
Fucus spp.	0	0	6.6
Phyllophora spp.	3.3	3.7	0
Laminaria sp.	0.3	0	0

Substrate.—The substrate of the spring spawning ground at Escuminac was dominated by sandstone rubble with patches of sand and by expanses of exposed bedrock colonized by algae. Eggs were not found on sandy bottom. Algal cover consisted mainly of Irish moss *Chondrus crispus*, with lesser amounts of rockweeks *Focus* spp. and leafweeds *Phyllophora* spp. (Table 2). The attachment of a high proportion of the egg to Irish moss was probably due to the widespread distribution of the plant rather than to a behavioral preference by Atlantic herring.

In 1980, all the spawn was found on algae. The absence of eggs at water depths below 4.3 m was due to the occurrence of sandy bottom and lack of vegetation. The large area of low egg density extending from the center of the bed (Figure 5) to the seaward edge of the surveyed area also coincided with an area of sandy bottom and sparse algal cover. Regression analysis showed significant correlations between egg density and abundances of rockweeds and leafweeds, but not of Irish moss (Table 3).

In 1981, estimates of mean algal cover ranged from 25 to 35%. A high percentage of eggs were attached to Irish moss. However, the correlation between egg density and macrophyte abundance was not significant in three cases out of four (Table 3). In 1983, 53.5% of the eggs in the samples were attached to macrophytes; of the remainder, 0.2% were attached to bare substrate and 46.3% were loose. The latter either were initially unattached or were dislodged during collection and subsequent handling of the samples. Regression analysis indicated a significant association between the number of attached eggs and the algal weights.

In 1984, more than 55% of the eggs at the first bed were attached to macrophytes while the per-

TABLE 3.—Regression analysis of Atlantic herring egg density (number/m^2 on algal abundance (wet weight/m^2 from the spring spawning surveys at Escuminac in Miramichi Bay, 1980–1984.

Year	Spawning bed	Marcophyte type	r	df	P
1980	1	*Chondrus*	0.40	32	NS
		Fucus	0.73	18	<0.01
		Phyllophora	0.77	17	<0.01
1981	1	*Chondrus*	0.16	29	NS
		Phyllophora	0.25	17	NS
	2	*Chondrus*	0.53	12	<0.05
		Phyllophora	0.24	3	NS
1983	1	Combined	0.55	34	<0.01
1984	1	Combined	0.33	38	<0.05
	2	Combined	0.06	15	NS
	3	Combined	0.01	4	NS

centages attached at the second and third beds were 0.8 and 29.7%, respectively (Table 4). A significant correlation between macrophyte biomass and egg abundance was obtained at bed one only (Table 3).

Density of egg deposition.—Densities of spring eggs deposited in Miramichi Bay varied widely from year to year (Table 5). On individual beds, densities ranged from 1,200 to 95,800 eggs/m^2. Total annual depositions in this spawning area varied from 1.5×10^7 to 1.1×10^{11} eggs among the four survey years.

Egg development.—About 70,000 Atlantic herring eggs were examined in the 1980 survey. All samples collected on May 13 were in developmental stages 3 and 4, and most samples collected May 20–22 were in stage 5 (Table 6). This indicated that the spawning bed was discovered a few days after spawning had started. Small quantities of herring spawn attached to algae drifted ashore after gale winds on May 10 and most of the eggs were in stage 2. The incubation period was estimated at 18 d, and the first spawn occurred on May 7 (Table 7).

In 1981, beach searches revealed small quantities of stage-1 and -2 eggs attached to Irish moss along a 100-m stretch of the beach on May 2, and stage-2 and -3 eggs were washed ashore by another gale on May 4. Among eggs collected by divers on May 5–6, 55% were in stage 3 (Table 6). Eggs collected on May 15 and 16 were in stages 4 and 5. The incubation period in 1981 was estimated at 20 d, and the first spawn occurred on April 29 (Table 7).

In 1983, egg samples were obtained from the discovery of the spawning bed on May 20 until June 1. On May 26–27, 75% of the eggs were in stages 3 and 4; by June 1, 97% of the eggs were in stage 5 (Table 6). Hatching appeared to be complete by June 3. The incubation period was estimated at 7 d, and the first spawn occurred on May 17 (Table 7). This was the first time that the actual duration of incubation could be determined with accuracy, the spawning bed having been found when the eggs were in stage 1.

In 1984, egg samples were collected May 12–31. Spawn from bed 3 indicated that it had been formed approximately 1–2 d after those on the other two beds. Egg development progressed from May 12 to 29 (Table 6). On May 31, large numbers of empty egg cases were noted, indicating hatching during the preceding night. Incubation period was estimated at 19 d, and the first spawn occurred on May 12 (Table 7).

Eggs extracted from stomachs of winter flounder also provided information on egg developmental stages (Figure 6). The temporal pattern of egg development from stomach analysis lagged about 1 d behind that from quadrat sampling of the spawning bed. The large decrease in the average

TABLE 4.—Percent distribution of Atlantic herring eggs by substrate type in Miramichi Bay, spring 1984.

Spawning bed	Percent egg deposition: On algae	On bare rock	Unattached
1	55.1	1.7	43.2
2	0.8	47.5	51.7
3	29.7	0	70.3

TABLE 5.—Atlantic herring egg densities (number/m^2) and total egg deposition on the spawning beds in 1980–1986, Gulf of St. Lawrence.

Year	Season	Spawning bed	Egg density (10^3 m^2)	Total egg deposition
		Miramichi Bay		
1980	Spring	1	8.6	7.5×10^8
1981	Spring	1	2.0	6.0×10^7
		2	1.2	1.5×10^7
1983	Spring	1	24.7	2.2×10^{10}
1984	Spring	1	85.9	4.8×10^{10}
		2	34.2	1.5×10^{10}
		3	95.8	1.1×10^{11}
		Fishermans Bank		
1985	Autumn	1	2,590.0	7.42×10^{11}
		2	250.0	1.16×10^{11}
		3	4.7	1.13×10^6
		4	7,940.0	1.96×10^{12}
		5	2,830.0	1.58×10^{12}
1986	Autumn	1	3,800.0	4.18×10^{12}

TABLE 6.—Percent composition of Atlantic herring eggs by developmental stage in 1980–1986, Gulf of St. Lawrence.

Date	Source of sample	Developmental stage of eggs 1	2	3	4	5
	Miramichi Bay					
1980						
May 10	Beach search		89	11		
May 13	Scuba diving			40	60	
May 20–22	Scuba diving				5	95
1981						
May 2–4	Beach search		75	25		
May 5–6	Scuba diving	7	28	55	10	
May 15–16	Scuba diving				43	57
1983						
May 20–23	Scuba diving	23	37	40		
May 26–27	Scuba diving		5	33	42	20
May 29–30	Scuba diving				11	89
Jun 1	Scuba diving				3	97
1984						
May 12–13	Predator stomachs	100				
May 15–17	Scuba diving	34	58	8		
May 21–23	and			45	55	
May 25–27	predator			5	75	20
May 28–29	stomachs				40	60
	Fishermans Bank					
1985						
Sep 9	Scuba diving	35	65			
Sep 13	Scuba diving				68	32
Sep 15	Scuba diving				10	90
1986						
Sep 6	Scuba diving		78	22		
Sep 7	Scuba diving		63	37		
Sep 9	Scuba diving				50	50

developmental stage on May 24 reflects the capture of two fish, one of which had eaten stage-1 eggs and the other a mixture of eggs in stages 2–4. These fish were collected in the area near Escuminac wharf, the one containing stage-1 eggs being captured outside of the surveyed spawning beds.

Natural mortality.—Low percentages of nonfertile eggs on the spawning beds indicated that egg mortalities were low (Table 8). In 1981, due to

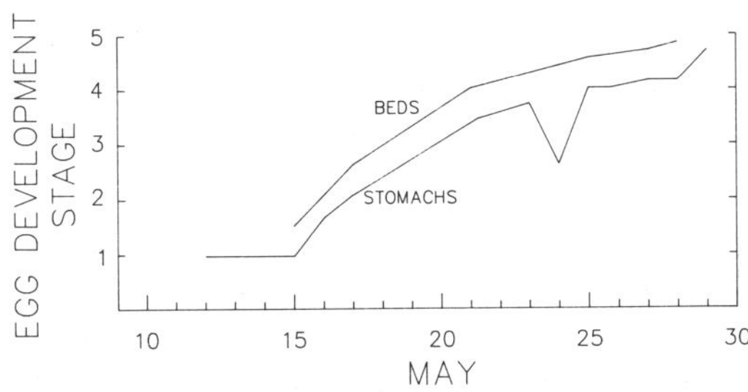

FIGURE 6.—Developmental stages of Atlantic herring eggs at predation sites on the spawning beds and in winter flounder stomachs, Miramichi Bay, May 1983.

TABLE 7.—Incubation periods of Atlantic herring eggs and water temperatures on the spawning beds, 1980–1986, Gulf of St. Lawrence.

Year	Estimated incubation period (d)	Calculated spawning date	Water temperature (°C)
		Miramichi Bay	
1980	18	May 7	6.9
1981	20	Apr 29	6.0
1983	17	May 17	7.5
1984	19	May 12	6.2
		Fishermans Bank	
1985	6	Sep 4	16.0
1986	7	Sep 3	14.8

the small size of the spawning beds and the light density of egg deposition, the percentage of nonfertile eggs was negligible. In 1984, natural mortality was markedly higher than in previous years. The highest egg mortality (9.4%) was on spawning bed 2 where algae were present at only 6 of 38 quadrat sample sites. The highest egg mortality was among loose eggs not attached to the substrate (9.6%); the lowest (1.2%) was for eggs attached to kelp *Laminaria* sp. and leafweeds (Table 9). Eggs attached to bare rock did not have a high natural mortality (4.7%). The high mortality on bed 2 may be due to the high percentage of loose eggs present.

Loss of eggs due to strong wave action and winds in 1984 was substantial. Field personnel noted algae and eggs washed ashore after strong onshore winds. The bulk of these eggs (mostly loose) and algae were found on the eastern side of the Escuminac wharf.

Predation on eggs.—Analysis of stomach contents of fish collected during the 4 years of surveys revealed that winter flounder were the major predators of eggs (Table 10). The mean number of eggs per stomach ranged between 1,527 in 1980 and 6,480 in 1984. Number of predators per square meter ranged between 0.29 to 1 for the same period. Stomachs of sculpin *Myoxocephalus* sp. contained 560–2,260 eggs, but these predators were much less abundant than winter flounder (Table 10). Stomachs of Atlantic tomcod *Microgadus tomcod* were only examined in 1981; they contained Atlantic herring eggs, but no counts were made.

Estimates of egg mortality due to winter flounder predation between spawning and quadrat sampling were calculated for the 1984 survey. The assumption was made that winter flounder feed once per day, based on stomach clearance rate

TABLE 8.—Natural mortality of Atlantic herring eggs on the spawning beds in spring 1980–1984, Miramichi Bay.

Year	Spawning bed	Percent of eggs nonfertile
1980	1	1.8
1981	1	0.5
	2	0.8
1983	1	3.9
1984	1	7.1
	2	9.4
	3	8.5

TABLE 9.—Natural mortality of Atlantic herring eggs by substrate type on the spawning beds in spring 1984, Miramichi Bay.

Substrate	Percent of eggs nonfertile
Epiphytes	6.1
Chondrus sp.	7.7
Fucus spp.	3.7
Phyllophora spp.	1.2
Laminaria sp.	1.2
Bare rock	4.7
None (eggs unattached)	9.6

(MacDonald et al. 1982) and on diurnal feeding periodicity (Frank and Leggett 1984). The 1-d lag in developmental stage between stomach and quadrat egg samples (Figure 6) supports the suggestion by MacDonald et al. (1982) that winter flounder clear 50% of the soft-tissued prey from their stomachs in 23–24 h.

All winter flounder were collected in daylight hours, and the mean number of herring eggs per stomach was considered to be representative of the daily mean egg consumption, given that winter flounder appear to be visually oriented predators (Olla et al. 1969). Estimated losses of Atlantic herring eggs to predation were 45%, 69%, and 47% for the three spawning beds, respectively.

Autumn Spawning Beds

1985 survey.—In 1985, five spawning beds were located on Fishermans Bank, which was surveyed from the end of August until the end of September (Table 1). The visibility was good (5–6 m), and currents occasionally were swift. Water temperature ranged from 15.0 to 15.9°C and from 15.1 to 17.0°C at the surface and bottom, respectively. The first bed (Figure 7) was located on September 6 at the west shoal of the Fishermans Bank. Its area was 286,000 m^2. The intensity of egg deposition was heavy (2.59×10^6 eggs/m^2) in most parts of the bed. Eggs formed a thick mat over an area of about 200 × 500 m. This mat frequently reached thicknesses of 20–30 layers of eggs (about 4.5 cm thick). Much of the egg mass on the spawning bed was covered with fungi, and some eggs were deteriorating. Microscopic examination of egg samples taken from various layers showed that most of the eggs were fertilized.

The second spawning bed (Figure 7) was located on September 9 about 500 m east of the first spawning bed. Eggs were deposited at depths of 10–15 m (MLW). Intensity of egg deposition (2.50×10^5 eggs/m^2) was less than at the first bed (Table 5). The estimated area of the spawning bed was 488,800 m^2.

The third spawning bed was found on September 13 about 1 km southeast of the second bed, in a water depth of 20 m. Spawning was light, and eggs were spread sporadically over the spawning

TABLE 10.—Predation by fish on Atlantic herring eggs on the spawning beds in the Gulf of St. Lawrence. P means present but not counted.

Year	Season	Species	Number of fish examined	Number of fish feeding on eggs	Number of eggs per stomach	Number of predators/m^2
			Miraimichi Bay			
1980	Spring	Winter flounder	30	15	1,527	1.00
1981	Spring	Winter flounder	45	19	1,709	P
		Atlantic tomcod	78	31	P	P
1983	Spring	Winter flounder	90	82	3,800	0.29
		Sculpin	36	17	2,260	0.02
1984	Spring	Winter flounder	83	81	6,480	0.83
		Sculpin	7	2	560	0.01
			Fishermans Bank			
1985	Autumn	Winter flounder	50	42	7,640	0.10
		Atlantic mackerel	40	36	3,920	P

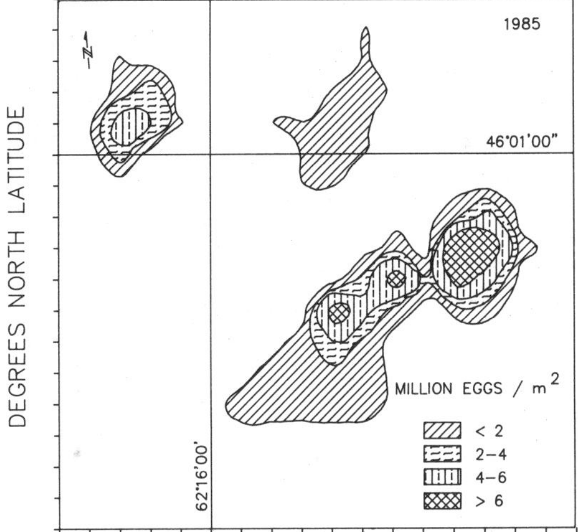

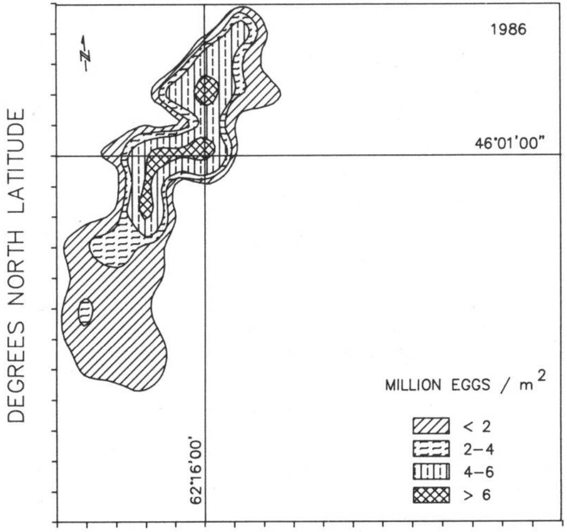

FIGURE 7.—Contour plots of Atlantic herring egg density on the spawning beds of Fishermans Bank, 1985–1986.

bed. Intensity of egg deposition ranged from 4,000 to 10,000 eggs/m^2, and averaged 4,714 eggs/m^2.

The fourth and fifth spawning beds were found on September 18, on either side of bed 3. By that time, all eggs from the previous spawn were hatched. Intensities of egg deposition on the two beds were 7.94 × 10^6 and 2.83 × 10^6, respectively (Table 5).

Examination of developmental stages (Table 6) showed a much faster rate of egg development in autumn than in spring. Incubation was estimated at 6 d (Table 7).

Samples from the first spawning bed showed an almost mass mortality of eggs. The eggs were solidly packed in a continuous mat with very few cavities between eggs or egg layers. Egg mass reached thicknesses up to 30 layers. Microscopic observations along cross sections of egg clusters indicated that most eggs (one average, 95%) had been fertilized and had reached some stage of development, mostly beyond the blastodisc stage (stage 1). The near-bottom layers had obviously died at an early stage (probably late blastodisc, onset of epiboly), whereas most of the eggs in the three top layers had advanced to embryonic stages that were close to the hatching stage.

Divers observed several fish predators on the spawning beds, including winter flounder, cunner *Tautogolabrus adspersus*, Atlantic cod *Gadus morhua*, pollock *Pollachius virens*, and Atlantic mackerel. Counts of benthic fish predators showed that winter flounder were most abundant (0.1 fish/m^2). Divers observed large schools of Atlantic mackerel feeding on Atlantic herring spawn. It was difficult to estimate the abundance of Atlantic mackerel because of their fast movement and general avoidance of divers.

Examination of stomach contents of fish collected showed that winter flounder and Atlantic mackerel were the major predators of Atlantic herring eggs. Mean numbers of eggs per stomach were 7,640 and 3,920 for the two species, respectively (Table 10).

1986 survey.—The 1986 search began on August 27 with spot dives, but no spawn was found. On August 31 and September 1, the complete bank was surveyed (74 stations) with underwater video. Again, no spawn was found, but schools of Atlantic herring were observed. On September 2, schools of Atlantic herring were again observed, and a large milt patch (about 400 m in diameter) discolored the water in the morning. Another survey of the bank in the afternoon of the same day revealed no trace of any egg deposition on the substrate.

On September 3, strong winds prevented divers from surveying the area. On September 4, a large heavy spawn was located by underwater video. Thirty stations were sampled, and the spawning bed perimeter was delineated. Another milt patch was located about 1 km southwest of the bank, where the water was 18 m deep, but no more spawn was found. The underwater camera survey was completed on September 5.

Sampling of the spawning bed by divers extended over 4 d (September 6–9). On September

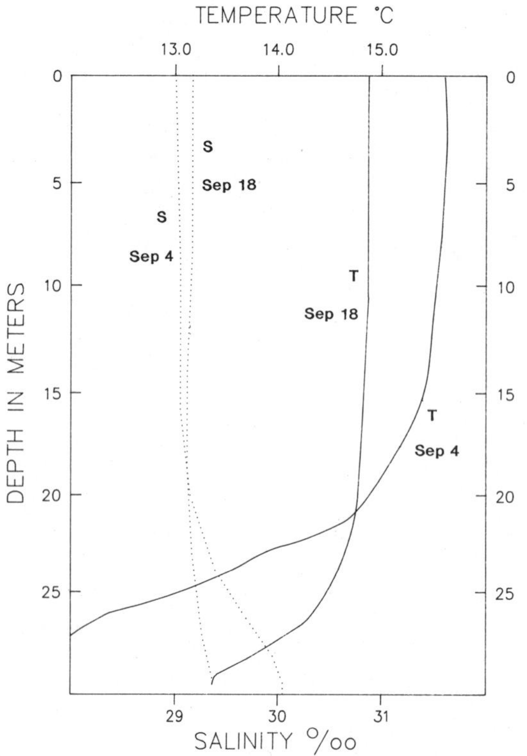

FIGURE 8.—Distributions of temperature (T) and salinity (S) near Fishermans Bank, September 1986.

18, a second survey of the bank was conducted but no other spawn was found, and the original bed was almost gone. Incubation period was estimated at 7–8 d. Temperature ranged between 14.0 and 16.3°C at the surface, and 14.0 and 15.8°C at the bottom. Off the bank, about 1 km from the spawning bed, surface temperature ranged between 14.8 and 15.6°C with a thermocline starting at 15 and 25 m depths (Figure 8).

The single bed was estimated to have an area of 1.1 km^2 (Figure 7) and an average egg density of 3.8×10^6 eggs/m^2 (Table 5). Egg deposition was thick, exceeding 5 cm in some areas. Average thickness of the spawn mat was 2 cm (20 layers of eggs). Some eggs were deteriorating, but no mass mortalities were observed.

Predation on Atlantic herring eggs was mainly by winter flounder and Atlantic mackerel. Video observation on September 5 also showed egg predation by cunners and by Atlantic herring themselves. Active cannibalism on egg lumps dislodged from the spawning bed was observed. It was not known whether the feeding Atlantic herring were the recently spent fish of the same spawning school or from a different school.

Spawning behavior.—The use of underwater video enabled spawning behavior to be observed and recorded in the southern Gulf of St. Lawrence for the first time. The observation of the milt on the bank prior to the deposition of eggs provided evidence that the females did not release their eggs until the milt was released. A few hours after the milt was discovered, the spawning school was seen moving in a highly organized manner about 30 cm above the bottom. The school was watched swimming in the milt for about 4 h while the vessel was still in the center of the milt patch. The vessel was allowed to drift at about 4 km/h so the substrate could be searched, but no spawn was detected during this time. To make sure that no spawn was deposited in other locations, the entire bank was surveyed with negative results. The deposition of eggs was observed the following day, and the spawning was complete.

For the first time during the spawning bed surveys, spawning by Atlantic herring was observed during daytime. Local fishermen who were interviewed did not recall any previous daytime spawning by this species.

Discussion

Aerial photographic surveys provided an easy and cost-effective method for locating herring gill nets on spawning grounds scattered along the shores of the Gulf of St. Lawrence. Comparison of the gill-net distributions during the 3-year study showed little change from year to year. The locations of the spawning grounds, as inferred from the aerial gill-net surveys, were similar to those in the late 1940s, although some variations in their size and relative importance had occurred. For example, the spring spawning grounds of the Caraquet and Magdalen islands were historically the largest in the Gulf of St. Lawrence. At present, the Caraquet spawning grounds are less important than those at Escuminac. The Magdalen Island spawning grounds disappeared after the collapse of the Atlantic herring stock there in the early 1970s.

The relative importance of the spring- and autumn-spawning components in the fishery has changed over the years. Nevertheless, the regularity of spawning times is clearly demonstrated by the seasonal distributions of gill-net catches since the 1940s. Until the mid-1960s, the herring catches were composed mostly of spring-spawning fish, and catches peaked in May. The autumn fishery expanded during the late 1960s and early 1970s, and catches were sometimes larger than those of the spring fishery. Spring

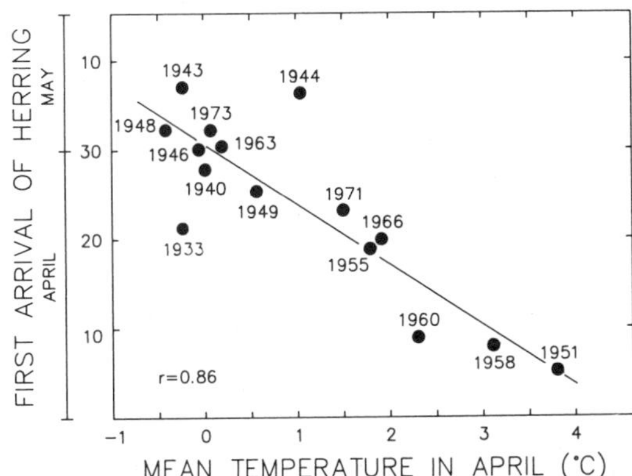

FIGURE 9.—Relationship between the time of first arrival of spring Atlantic herring spawners and surface water temperature on the spawning grounds at Magdalen Islands, from sporadic observations during 1933–1973. (From Messieh 1977.)

spawners arrive on the spawning grounds in late April or early May and spawning is over by the end of May or early June. Autumn spawners arrive on the spawning grounds in late July or early August and stay until the end of September.

The situation is similar to that of 135 years ago. Perley (1850) reported that "the common herrings appear in the Gulf of St. Lawrence at the end of April, or early May, and the fishery continues until about 10th June, when they retire to deep water, having deposited their spawn. Another herring appears on the coast about the 20th August and remains inshore for a month. . . ." The ecological significance of the multiplicity of herring spawning locations and spawning times, in relation to larval production, has been discussed by several authors (e.g., Cushing [1975] and Blaxter and Hunter [1982] for the northeast Atlantic herring; Graham 1982, Lambert [1984], and Lambert and Ware [1984] for the northwest Atlantic herring; Hay [1985] for the Pacific herring).

The role of temperature in influencing the arrival time on spawning grounds was demonstrated by Messieh (1977), who reported significant correlations between these variables. He showed a linear relationship between the arrival time of Atlantic herring on spawning grounds near the Magdalen Islands and mean surface temperature over the year 1933–1973 (Figure 9). Spawning occurred in late April and early May of cold years and in early April of warm years.

The effect of water temperature on fish production and recruitment is well documented (see Templeman and Hodder 1965; Templeman 1972; Cushing 1982). Cushing (1982) discussed the different explanations for the mechanism by which temperature affects populations and reported that all correlations of population characteristics with temperature are unlikely to be explained in the same way.

Grimm (1983) found a relationship between the time and location of Atlantic herring spawning and bottom temperature in the Georges Bank and Nantucket Shoals area. Spawning occurred on the northern part of Georges Bank from September to November in 1971–1973, but it did not begin until October in 1974–1976. On Nantucket Shoals, spawning usually occurred from October to December. After 1975, the onset of spawning was delayed and the spawning season was shorter in both areas. These changes coincided with a general increase in water temperature in the Georges Bank–Gulf of Maine region from the late 1960s to 1977. The large volume of warm water over the spawning beds on Georges Bank in these years may have affected spawning, hatching success, or both. On the Nantucket Shoals spawning grounds, mean bottom temperatures were lower than those on Georges Bank, particularly in 1976 and 1977, when newly hatched larvae were prevalent on Nantucket Shoals but not on Georges Bank.

The spring spawning bed substrates in the Gulf of St. Lawrence are mainly sandstone, rubble, and expanses of exposed bedrock colonized by algae. Algal cover is dominated by Irish moss, rockweeds, leafweeds, and kelp at depths of 0.9–6.0 m. The large percentage of attached eggs found on Irish moss may have been a function of the abundance and wide-spread distribution of this plant in the spawning areas, rather than of a preference for this species by Atlantic herring. The finding of spring spawn on bare rock in 1984 contrasted with observations in previous years with observations made by Tibbo et al. (1963) in the Caraquet area. Spawning on bare rock is common in the northeastern Atlantic herring stocks (Parrish et al. 1959; Dempsey and Bamber 1983).

The occurrence of spring herring spawning beds in shallow waters of the littoral zone often results in eggs being washed ashore during gales. Prince (1907) stated that, in some years, herring eggs were heaped knee high for kilometers along the New Brunswick shore after severe gales. Escuminac-area fishermen reported that enough eggs were washed ashore to be collected for fertilizer as recently as the early 1960s. The negligible amounts of beached material during our surveys from 1980 to 1984 most likely reflected the decline in herring stocks and the small amount of spawn compared to previous years.

The physical characteristics of the autumn spawning beds in the Gulf of St. Lawrence are different than those of the spring beds. The autumn beds are in deeper waters (10–25 m) and further offshore than the spring ones. In some areas such as Caraquet, Kouchibouguac, and North Point, the spring and autumn beds are in the same general area, although the latter are in deeper waters.

The Fishermans Bank autumn spawning bed is presently the most important bed in the Gulf of St. Lawrence. The bank is almost flat with a depth range of 10–25 m. It consists of bedrock with cobble and negligible vegetation. Spawning occurs over the whole depth range, but eggs are most concentrated at depths of 15–20 m depth. Spawning appears to start on the west shoal of the bank and then progresses eastward. This progression agrees well with the general route of Atlantic herring movement observed during the spawning migration.

The physical characteristics of the spawning beds on Fishermans Bank are similar to those on Georges Bank observed by Drapeau (1973) and Caddy and Iles (1973) in the early 1970s. The only exception is that Georges Bank is much deeper (average, 40 m). In both locations, spawn was observed on gravel patches devoid of sand, and the tidal currents are relatively intense (1.0 m/s and 1.2 m/s for Georges Bank and Fishermans Bank, respectively). It appears that Atlantic herring choose a gravel or bedrock substrate to ensure stability of deposited eggs. The high-energy environment of spawning beds with strong currents prevents silt accumulation that could smother the eggs and provides better circulation to supply oxygen and remove the metabolites (Hempel 1971; H. Rosenthal, Biologische Anstalt Helgoland, Hamburg, Federal Republic of Germany, personal communication).

Bowers (1980) reviewed the characteristics of northeastern Atlantic herring spawning grounds and reported a wide range of substrates, ranging from coarse sand to rock. However, gravel was found always on at least part of the spawning ground. Observations of egg and sediment distribution in Milford Haven spawning ground, southwest Wales (Clarke and King 1985), showed the heaviest spawning on substrates that were half coarse shell and half stone gravel 2–5 mm in diameter. Shell was composed of broken pieces of topshell, oysters, and other bivalves. Finer shell gravel (<1.5 mm) and larger stones (up to 20 cm) also received some spawn, though this was less dense.

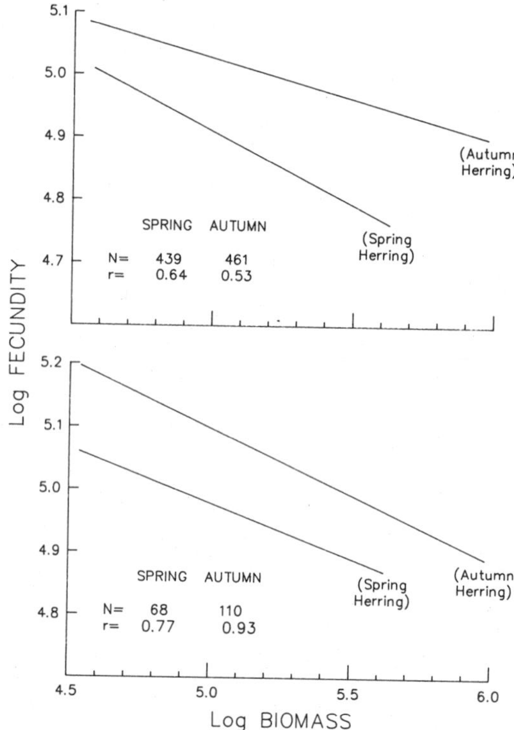

FIGURE 10.—Regression of $\log_{10}$ fecundity against $\log_{10}$ biomass of spring- and autumn-spawning Atlantic herring in the southern Gulf of St. Lawrence. The upper panel shows sizes combined; the lower panel shows 300–340-g fish only.

One of the major differences between spring-spawning and autumn-spawning Atlantic herring in the Gulf of St. Lawrence is in the density of their deposited eggs. The greatest spring density measured was 9.6×10^4 eggs/m^2, whereas autumn densities reached 3.8×10^6 eggs/m^2 (Table 5). The higher egg density on autumn beds could be due partly to a larger spawning-stock biomass, and partly to a higher spawner fecundity (Figure 10). Messieh (1976) found that the autumn-spawning Atlantic herring in the 31-cm length group (about 5 years old) produce 112,900 eggs per female, compared with 69,700 eggs per female for the same length group of spring spawners.

Mass mortalities of Atlantic herring eggs were observed on Fishermans Bank in autumn 1985. The observations that large areas of the spawning beds were covered with fungus led to the investigation of a possible relationship between fungal

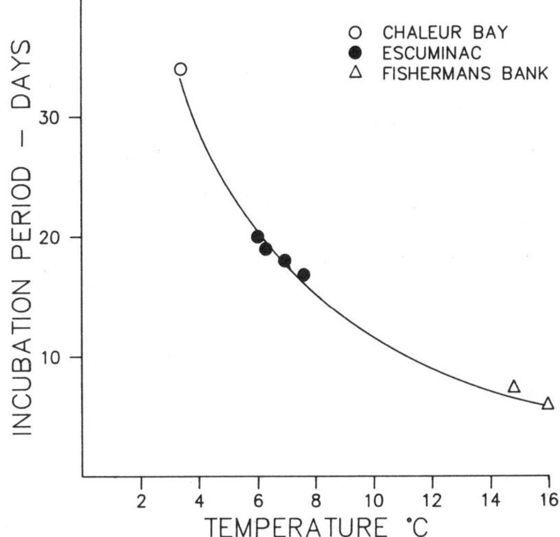

FIGURE 11.—Relationship between water temperature and incubation period of Atlantic herring eggs in the Gulf of St. Lawrence.

infection and mass mortality of eggs. However, the available information from microscopic examination indicated that fungal growth on the spawning beds was a secondary effect, following the mass mortality and deterioration of several extended egg batches that had died from other causes. These results indicated that the heavy egg deposition, forming a thick mat of 20 to 30 layers of eggs, did not permit enough water circulation to deliver oxygen, remove metabolites, and sustain egg development and survival (Messieh and Rosenthal 1986). High density of herring egg deposition and high mortality have been reported in other areas (e.g., Rosenthal and Alderdice 1976 for Pacific herring).

Temperature was inversely related to the incubation period of Atlantic herring eggs (Figure 11). High temperatures accelerate and low temperatures retard development of teleost embryos (Parrish and Saville 1965). Blaxter (1956) showed this relationship between the rate of development and temperature from a regression of the type $(T-T_0)(D-D_0) = K$; D is the time to reach a given stage of development at temperature T, T_0 is the temperature at which development takes an infinite time, D_0 is the time to hatching at infinite temperature, and K is a constant.

In the Gulf of St. Lawrence, spring temperatures that averaged between 6.0 and 7.5°C resulted in incubation periods of 20 d to 17 d, respectively. In autumn, the bottom temperature ranged between 15.0 and 16.0°C, and the incubation periods were 8 and 6 d, respectively. Autumn temperatures of 17.5°C observed during spawning in the Gulf of St. Lawrence appear to be near the optimal temperature known for Atlantic herring (Blaxter and Hunter 1982), whereas spring temperatures of 3.7°C in the same area (Jean 1956) are near the lower end of spawning temperature.

Winter flounder were the major predators on Atlantic herring eggs in spring, and winter flounder and Atlantic mackerel were the major predators in autumn. Of all fish predation on Atlantic herring eggs, that by winter flounder was the heaviest in all years of the present study. Winter flounder of all sizes could be seen everywhere with distended abdomens, camouflaged on the spawning beds. Their stomachs contained nothing but Atlantic herring eggs. The winter flounder spawning season coincided with that of Atlantic herring. Larval distributions of the two species in spring in Miramichi Bay were almost identical (Figure 12). This similarity indicates a complex relationship between the winter flounder and Atlantic herring that involves predation at the adult stage and competition at the larval stage.

The discovery of feeding by Atlantic mackerel on Atlantic herring eggs is of special ecological interest because of the possible interrelation between these species as hypothesized by Lett and Kohler (1976), Winters (1976), and Skud (1982). This is the first time, to our knowledge, that Atlantic mackerel (pelagic feeders) have been reported to prey on herring eggs (demersal). Divers observed Atlantic mackerel actively feeding on lumps of Atlantic herring eggs loosened from substrate. Examination of Atlantic mackerel stomachs from the spawning bed showed that they were full of Atlantic herring eggs. Skud and Messieh (unpublished) tested the relationship between the abundances of these species in the Gulf of St. Lawrence in light of the environmental variations and changes in population dominance. Atlantic herring and Atlantic mackerel have alternated as dominant and subordinate species in the pelagic biomass in the Gulf of Maine and the Gulf of St. Lawrence. The dominance of Atlantic mackerel in the Gulf of St. Lawrence coincided with the decline of the Atlantic herring population in this area and the increasing trend in water temperature since the early 1970s.

Predation rates varied between 45 and 69% of the initial Atlantic herring spawn in 1984. These estimates did not include invertebrate predators such as lobsters and starfish. Predation rates in 1980

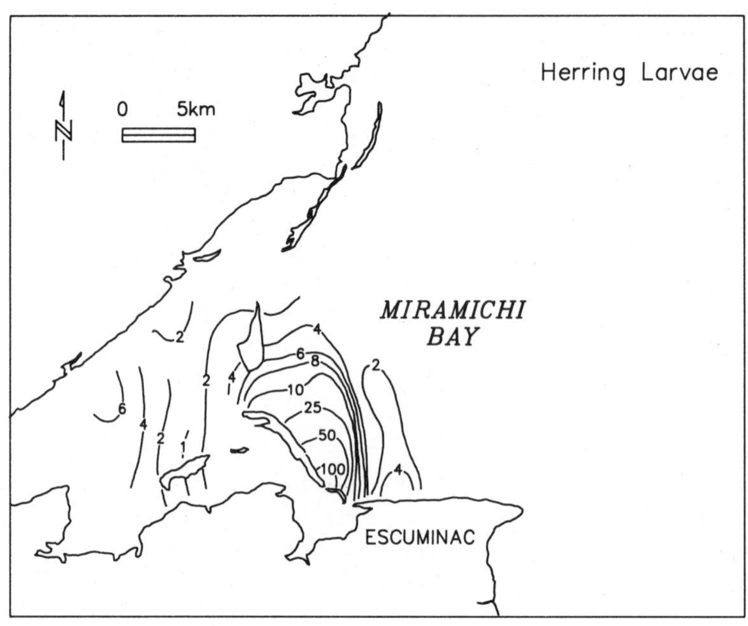

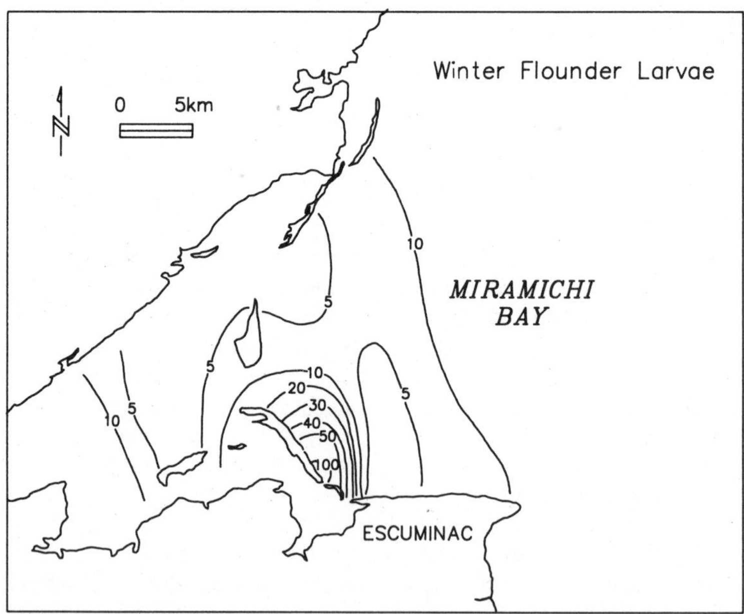

FIGURE 12.—Distribution of Atlantic herring larvae and winter flounder larvae in Miramichi Bay, May 1980. Isopleths indicate numbers of larvae/10^3 m^3, determined from an ichthyoplankton survey.

and 1981 were much less than in 1984, most likely due to the small Atlantic herring stock biomass and light egg deposition in those years. Estimates of predation rates reported in this study are similar to estimates in other areas. Caddy and Iles (1973) estimated that approximately 8% of the spawn on Georges Bank had been removed by predation within 1–2 d of spawning. Hardwick (1973) reported predation rates on Pacific herring eggs of 56–99%, and birds accounted for the major predation, which took place within 3 d of spawning.

Little information is available on Atlantic herring spawning behavior on either side of the north Atlantic Ocean (Blaxter and Hunter 1982). Most of what is known comes from observation of fish in captivity. Observations on the spawning behavior of Atlantic herring reported in the present study are of special interest because of the controversy surrounding this topic. Haegele and Schweigert (1985) reviewed this subject and cited Ewart (1884), who wrote that the female Atlantic herring stops, almost touching the substrate, and releases a ribbon of eggs that adheres to the bottom. Several males form a circle around and 20–25 cm above her. They release short ribbons of milt 1–2 cm long, which slowly settle toward the bottom in the vicinity of the ova. Ewart also suggested that females spawn for 3 d, whereas the males take about twice that long. Holliday (1960) confirmed the observations on spawning behavior based on captive fish.

Only one in situ observation on spawning behavior has been reported for Baltic herring *Clupea harengus membras*. Aneer et al. (1983) indicated an almost random intermingling of males and females, quivering and yawning, on the spawning grounds. For Pacific herring, Hay (1985), observed small patches of milt-discolored water, but no egg deposition. In another field study in Barkley Sound, on the west coast of Vancouver Island, Hay took small beach-seine samples on the spawning grounds as the fish were spawning. Almost all of the males were totally spent, but most females were only partially spent. These observations suggest that males probably initiate the spawning act, after which females deposit eggs on the substrate. On the other hand, Hourston et al. (1977) noted different behavioral patterns for captive Pacific herring, including substrate testing and substrate spawning, both of which involve contact with the urogenital papilla and substrate. Hay (1985) reported that it is difficult to understand how visual stimuli can operate once the spawning act has begun. In very heavy spawnings, the number of egg layers deposited increases so that most eggs are deposited on other eggs. The significance of this is that if females were to use tactile sensory information to assess the substrate, the only surface available would be that of previously deposited eggs.

The in situ observations of spawning behavior on Fishermans Bank, reported in the present study, provide evidence that female Atlantic herring do not release their eggs until the males have released their milt. This sequence in the spawning act would ensure proper fertilization of eggs before they adhere to the substrate. Messieh and Rosenthal's (1986) observations that almost all eggs were fertilized even in deposits of that were 30 layers deep, provide additional evidence for this spawning behavior. If the females were to deposit their eggs prior to milt release, one would expect a large percentage of nonfertile eggs, particularly in the inner layers of the egg mat.

References

Aneer, G., G. Florell, U. Kautsky, S. Nellbring, and L. Sjostedt. 1983. *In situ* observations of Baltic herring (*Clupea harengus membras*) spawning behaviour in the Asko-Landsort area, northern Baltic proper. Marine Biology (Berlin) 74:105–110.

Baxter, I. G. 1971. Development rates and mortalities in Clyde herring eggs. Rapports et Procès-Verbaux des Réunions Conseil International pour l'Exploration de la Mer 160:27–29.

Blaxter, J. H. S. 1956. Herring rearing—II. The effect of temperature and other factors on development. Marine Research Department of Agriculture and Fisheries for Scotland 5.

Blaxter, J. H. S., and J. R. Hunter. 1982. The biology of clupeoid fishes. Advances in Marine Biology 20: 1–120.

Bowers, A. B. 1980. Characteristics of herring spawning grounds. International Council for the Exploration of the Sea, C.M. 1980/H:13, Copenhagen.

Caddy, J. F., and T. D. Iles. 1973. Underwater observations on herring spawning grounds on Georges Bank. International Commission for the Northwest Atlantic Fisheries, Research Bulletin 10:131–139.

Clarke, D. R., and P. E. King. 1985. Spawning of herring in Milford Haven. Journal of the Marine Biological Association of the United Kingdom 65:629–639.

Cushing, D. H. 1975. Marine ecology and fisheries. Cambridge University Press, London.

Cushing, D. H. 1982. Climate and fisheries. Academic Press, London.

Dempsey, C. H., and R. N. Bamber. 1983. Spawning of herring (*Clupea harengus* L.) in the Blackwater Estuary, spring 1979. Journal du Conseil, Conseil International pour l'Exploration de la Mer 41:85–92.

Drapeau, G. 1973. Sedimentology of herring spawning grounds on Georges Bank. International Commission for the Northwest Atlantic Fisheries, Research Bulletin 10:151–162.

Ewart, J. C. 1884. Natural history of the herring. Report of The Fishery Board of Scotland 2:61–73.

Frank, K., and W. C. Leggett. 1984. Selective exploitation of capelin (*Mallotus villosus*) eggs by winter flounder (*Pseudopleuronectes americanus*): capelin egg mortality rates, and contribution of egg energy to the annual growth of flounder. Canadian Journal of Fisheries and Aquatic Science 41:1294–1302.

Graham, J. J. 1982. Production of larval herring (*Clupea harengus*) along the Maine coast, 1964–78. Journal of Northwest Atlantic Fishery Science 3:63–85.

Grimm, S. K. 1983. Changes in time and location of

herring (*Clupea harengus* L.) spawning relative to bottom temperature in the Georges Bank and Nantucket Shoals areas, 1971–77. Northwest Atlantic Fisheries Organization Scientific Council Studies 6: 15–34.

Haegele, C. W., and J. F. Schweigert. 1985. Distribution and characteristics of herring spawning grounds and description of spawning behaviour. Canadian Journal of Fisheries and Aquatic Sciences 42 (supplement 1):39–55.

Hardwick, J. E. 1973. Biomass estimate of spawning herring (*Clupea harengus pallasi*) herring eggs and associated vegetation in Tomales Bay. California Fish and Game 59:36–61.

Hay, D. E. 1985. Reproductive biology of Pacific herring (*Clupea harengus pallasi*). Canadian Journal of Fisheries and Aquatic Sciences 42 (supplement 1): 111–126.

Hempel, G. 1971. Egg production and egg mortality in herring. Rapports et Procès-Verbaux des Réunions Conseil International pour l'Exploration de la Mer 160:8–11.

Holliday, F. G. T. 1960. The control of spawning in the herring. International Council for the Exploration of the Sea, C.M. 1960/H:38, Copenhagen.

Hourston, A. S., H. Rosenthal, and N. Stacy. 1977. Observations on spawning behaviour of Pacific herring in captivity. Meeresforschung 25:156–162.

Humphreys, R. D., and A. S. Hourston. 1978. British Columbia herring spawn deposition survey manual. Canada Fisheries and Marine Service Miscellaneous Special Publication 38.

Jean, Y. 1956. A study of spring and fall spawning herring (*Clupea harengus* L.) at Grande-Rivière, Bay of Chaleur, Quebec. Quebec Department of the Fisheries, Contribution 49, Québec, Canada.

Lambert, T. C. 1984. Larval cohort succession in herring (*Clupea harengus*) and capelin (*Mallotus villosus*). Canadian Journal of Fisheries and Aquatic Sciences 41:1552–1564.

Lambert, T. C., and D. M. Ware. 1984. Reproductive strategies of demersal and pelagic spawning fish. Canadian Journal of Fisheries and Aquatic Sciences 41:1565–1569.

Lett, P. F., and A. C. Kohler. 1976. Recruitment: a problem of multispecies interaction and environmental perturbations, with special reference to Gulf of St. Lawrence Atlantic herring. Journal of the Fisheries Research Board of Canada 33:1353–1371.

MacDonald, J. S., K. G. Waiwood, and R. H. Green. 1982. Rates of digestion of different prey in Atlantic cod (*Gadus morhua*), ocean pout (*Macrozoarces americanus*), winter flounder (*Pseudopleuronectes americanus*) and American plaice (*Hippoglossoides platessoides*). Canadian Journal of Fisheries and Aquatic Sciences 39:651–659.

Messieh, S. 1976. Fecundity studies on Atlantic herring from the southern Gulf of St. Lawrence and along the Nova Scotia coast. Transactions of the American Fisheries Society 105:384–394.

Messieh, S. 1977. The regularity of spawning time of Atlantic herring in the Gulf of St. Lawrence. International Council for the Exploration of the Sea, C.M. 1977/H:25, Copenhagen.

Messieh, S., and R. Pottle. 1986. Distribution of gillnets and spring spawning grounds of Atlantic herring in the southern Gulf of St. Lawrence as inferred from aerial surveys. Canadian Technical Report of Fisheries and Aquatic Sciences 1504.

Messieh, S., R. Pottle, P. MacPherson, and T. Hurlbut. 1985. Spawning and exploitation of Atlantic herring (*Clupea harengus*) at Escuminac in the southwestern Gulf of St. Lawrence, spring 1983. Journal of Northwest Atlantic Fishery Science 6:125–133.

Messieh, S., and H. Rosenthal. 1986. Mass mortality of herring eggs on spawning beds on and near Fishermans Bank, Gulf of St. Lawrence (NAFO Div. 4T) during fall 1985. International Council for the Exploration of the Sea, C.M. 1986/H:4, Copenhagen.

Olla, B., R. Wicklund, and S. Wilk. 1969. Behavior of winter flounder in a natural habitat. Transactions of the American Fisheries Society 98:717–720.

Parrish, B. B., and A. Saville. 1965. The biology of the northeast Atlantic herring populations. Oceanography and Marine Biology: An Annual Review 3:323–373.

Parrish, B. B., A. Saville, R. Craig, I. G. Baxter, and R. Preistley. 1959. Observations on herring spawning and larvae distribution in the Firth of Clyde in 1958. Journal of the Marine Biological Association of the United Kingdom 38:445–453.

Perley, M. H. 1850. Reports on the sea and river fisheries of New Brunswick. J. Simpson, Fredericton, Canada.

Prince, E. E. 1907. The eggs and early life history of the herring, gaspereau, shad and other clupeoides. Contributions of the Canadian Biological Board 11: 95–110.

Rosenthal, H., and D. F. Alderdice. 1976. Sublethal effects of environmental stressors, natural and pollutional, on marine fish eggs and larvae. Journal of the Fisheries Research Board of Canada 33:2047–2065.

Skud, B. E. 1982. Dominance in fishes: the relationship between environment and abundance. Science (Washington, D.C.) 216:144–149.

Templeman, W. 1972. Year-class success in some North Atlantic stocks of cod and haddock. International Commission for the Northwest Atlantic Fisheries Special Publication 8:223–239.

Templeman, W., and V. M. Hodder. 1965. Distribution of haddock on the Grand Bank in relation to season, depth and temperature. International Commission for the Northwest Atlantic Fisheries Special Publication 6:171–187.

Tibbo, S. N., D. J. Scarratt, and P. G. McMullon. 1963. An investigation of herring (*Clupea harengus* L.) spawning using free diving techniques. Journal of the Fisheries Research Board of Canada 20:1067–1079.

Winters, G. H. 1976. Recruitment mechanisms of southern Gulf of St. Lawrence Atlantic herring (*Clupea harengus harengus*). Journal of the Fisheries Research Board of Canada 33:1751–1763.

American Fisheries Society Symposium 5:49–59, 1988

Transplanting Pacific Herring Eggs in British Columbia: A Stocking Experiment

D. E. Hay

Department of Fisheries and Oceans, Fisheries Research Branch, Pacific Biological Station Nanaimo, British Columbia V9R 5K6, Canada

J. B. Marliave

Vancouver Public Aquarium, Post Office Box 3232 Vancouver, British Columbia V6B 3X8, Canada

Abstract.—Spring storms can dislodge eggs of Pacific herring *Clupea harengus pallasi* spawned in intertidal and shallow subtidal locations. Eggs sometimes pile up in drifts on the shore but remain viable for several days or longer. The eggs could be moved without damage if they were kept damp but out of water. Immersion in nonflowing seawater led to complete mortality. In 1982 and 1984, 17,000 kg and 9,000 kg, respectively, of wind-drifted eggs were transplanted in an attempt to establish Pacific herring in a new location. Many eggs survived to hatching and surveys of larvae were made in the new location. Length and weight analyses showed that larvae fed and grew. Catches of larvae decreased within 3 weeks after hatching; after 4 weeks, abundance was so low that sampling was stopped. If some fish had survived to sexual maturity, spawning would have been expected in 1985–1987, but none occurred. Although a new spawning population was not started, the work led to development of techniques useful for future transplantation attempts.

This paper reports on an experimental attempt to restock an area with eggs of Pacific herring *Clupea harengus pallasi* transplanted from a different natural spawning site. The possibility of transplanting Pacific herring eggs has wide interest and appeal for managers because of its potential applications. For instance, if spawn could be moved successfully, managers could use this technique to increase production, as is done in salmon hatcheries (Moser and Hansen 1986). Also, spawn transplants could alleviate the effects of spawning habitat degradation from nearshore industrial or urban development. The concept also has important biological implications because the successful establishment of new stocks or runs would improve our understanding of the biological basis of herring stocks and would affect theories of larval retention (Iles and Sinclair 1982) and homing (Hourston 1982).

In this paper we also provide some biological information on spawning history in the vicinity of the Pacific herring transplants. Then, from the results of this work, we discuss the criteria for assessing the success or failure of Pacific herring transplants.

Methods

Study area and source of eggs.—In 1982 and 1984, Pacific herring eggs were transplanted from wind drifts or windrows that sometimes form following strong storms in March and April during the herring spawning season. The times and locations of wind drifts cannot be predicted; in many years, they do not occur at all. No substantial amounts of wind-drifted eggs were found in 1983, so no transplants were made that year.—In each year of the study, we searched likely wind-drift locations from aircraft. Local residents near waterfronts and federal fishery officers also reported incidences of wind-drifted spawn to us. Spawn was recovered in 1982 and 1984 from several sites where spawning had occurred during the previous 10 d (Figure 1). At specific sites, spawning usually lasts 1–3 d (Hay and Kronlund 1987), so most wind-drifted eggs at a site were nearly the same in age and development.

The eggs were transplanted to several intertidal locations around the perimeter of Hidden Basin, an indentation of Nelson Island (Figure 2). We chose Hidden Basin because Pacific herring had previously spawned in the immediate vicinity, but not in recent years. Further, Tidal Rush Marine Farms, a company that participated in the project, was located within the basin and was well placed to monitor eggs and larvae. Hidden Basin is about 1.5 km long, 0.5 km wide, and 40 m deep in the middle; its area, about 850,000 m^2, was estimated from a nautical chart. The natural deposition of

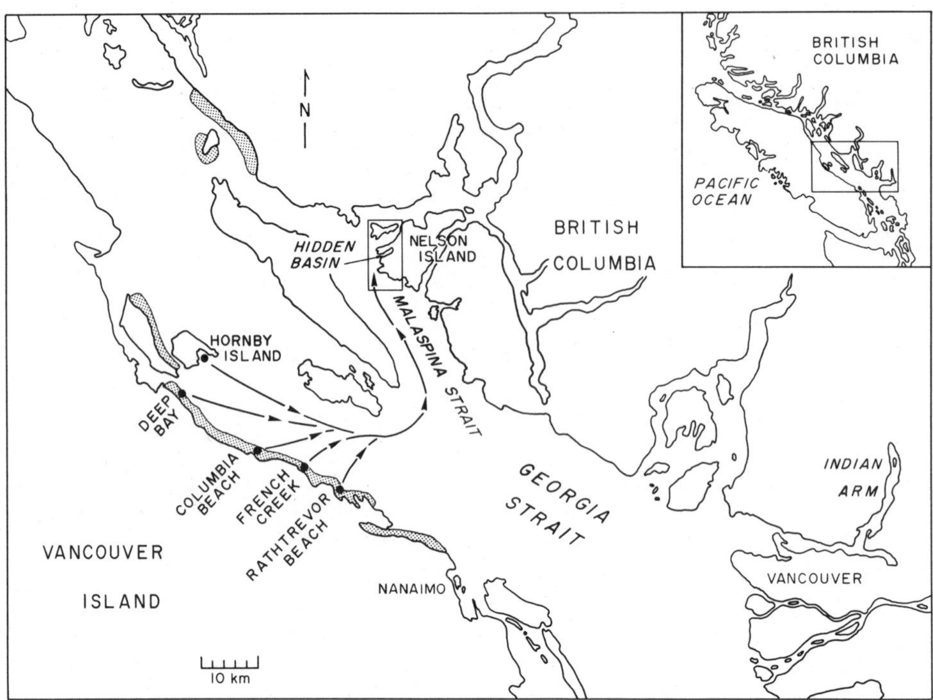

FIGURE 1.—Locations of wind-drifted spawns of Pacific herring (solid circles) in 1982 and 1984 and the transplant location, Hidden Basin, on Nelson Island. Stippling along the shoreline indicates locations of large spawning sites regularly used by Pacific herring.

Pacific herring spawn usually does not exceed depths of 10 m in southern British Columbia (Haegele et al. 1981), and we scattered most of the transplanted eggs among submerged vegetation at depths of 1–5 m. The density of these eggs probably was less than that of a natural spawn.

The history of spawning within Hidden Basin was deduced from spawning records made annually by federal fisheries officers. The total amount of spawn, estimated in terms of a spawn index developed by Hay and Kronlund (1987), was used to assess the frequency and magnitude of spawnings from 1928 to 1986 in the immediate vicinity of the transplant locations.

Egg viability.—Wind-drifted spawn was examined under a dissecting microscope on the beach, prior to removal. In most cases, the proportion of dead or moribund eggs was estimated after examination of random samples of at least 500 eggs. Transplants were not made if most of the eggs were dead or moribund, or if they were within 1–2 d of hatching; dead eggs were opaque, live eggs were clear. Moribund eggs, containing embryos whose development had been arrested, usually were slightly less transparent than healthy eggs. For most transplants, 200–300-g subsamples of wind-drifted eggs were taken to the laboratory at the Pacific Biological Station in Nanaimo (Figure 1) for analysis of the relative proportions, by weight, of egg and nonegg material. Eggs were separated individually from vegetation and debris, which often consisted of macrophytes, waterlogged wood debris, and sand pebbles. The total weights of egg and nonegg fractions were determined to the nearest 0.1 g.

Transport experiments.—Several different methods of transporting wind-drifted spawn were attempted. In all instances the eggs were removed from the beach manually. Most of the wind-drifted spawn was loaded onto trucks parked close to the beach site, usually within 500 m. Then the trucks were driven to a marina at French Creek, where eggs were loaded onto a transport vessel (a 10-m-long landing craft) that moved the spawn directly to Hidden Basin. In some instances it was possible to load spawn directly from the beach onto the landing craft, which could be moored close to shore on a rising tide if the beach was not too rocky. Sometimes loading the vessel took 6–12 h; moving the vessel to Hidden Basin took as much as 12 h.

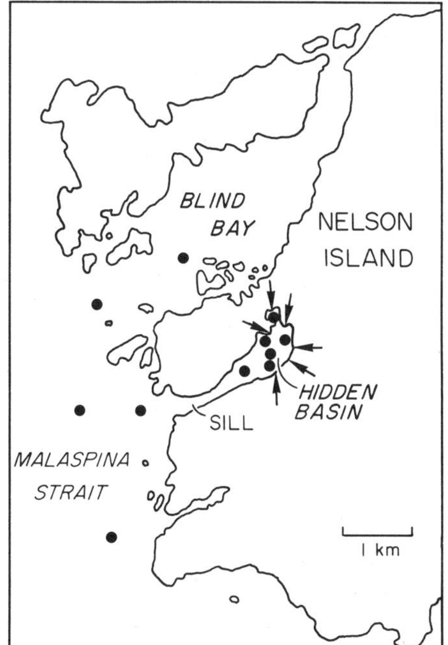

FIGURE 2.—Transplant sites (arrows) for Pacific herring spawn in Hidden Basin, western Nelson Island, British Columbia. The solid circles indicate the larval sampling sites in Hidden Basin and the waters outside it.

The initial transplants were made on March 15, 1982, by filling burlap sacks with about 30 kg of wind-drifted eggs and other beach debris. The eggs remained in the sacks without cooling until they arrived at the transplant site. The next transplant, on March 17, 1982, was smaller, and two different methods were tried: (1) about half the eggs were piled loosely on the deck, covered with burlap, and periodically splashed with water; (2) the other half was held in nonflowing water in rigid plastic containers of about 200-L capacity. These containers were half filled with loosely packed eggs and then filled to the top with seawater. In all subsequent transplants in 1982 and 1984, the eggs were loaded directly onto the vessel and piled loosely, similar to the way they were found on the beach. When the boat was running, its bow trim was adjusted so the eggs were intermittently sprayed with seawater. This kept the eggs cool and moist but not submerged.

Estimating the number of eggs transplanted.—The total weight of transplanted material was estimated in kilograms. This was adjusted by the proportion of eggs in each transplant. The resulting total weight of transplanted eggs was divided by 2.38 mg, the mean wet weight of one egg (SE, 0.34: Hay and Miller 1982). This gave the number of transported eggs. The corresponding biomass of the spawning stock was estimated by dividing the egg number by 10^8, the approximate number of eggs produced by 1,000 kg of sexually mature Pacific herring with a 1:1 sex ratio (Hay 1985).

Egg survival during transport.—Several experiments were conducted to determine the best way to maintain eggs out of water. In the experiment on March 17 (the date of the second transplant), a few kilograms of wind-drifted eggs from Rathtrevor Beach were taken directly to the laboratory at the Vancouver Public Aquarium. Two incubation tests were conducted, each with about 10,000 eggs. Each group of eggs was placed on a Vexar screen, supported over a water tray but out of the water, and covered with a damp paper towel for incubation. One tray had an air stone that provided a continuous seawater spray, the other did not. About 1,000 eggs from each treatment were examined daily for viability. The air temperature varied from 15 to 18°C. In the second experiment, incubation tests were made on the beach at a marina in French Creek. Eggs were held in six perforated wooden crates, each with a volume of about 1 m^3. Three of the crates were sprayed with seawater, which provided a within-crate flow rate of about 1 L/min; the remaining crates were not sprayed. Within each treatment (spraying, no spraying), eggs were loaded to a depth (thickness) of 0.3 m in one crate, 0.6 m in a second, and 1.0 m in the third. The experiment was set up on March 29–30 with eggs from the Columbia Beach wind-drift site. Egg survival was monitored from April 1 to April 10 with samples of 500–1,000 eggs taken from a depth of 15 cm in each crate.

Monitoring egg and larval survival.—In Hidden Basin, samples of incubating eggs were monitored frequently, and egg viability was estimated 1–2 d before hatching. Larvae were monitored by ichthyoplankton surveys with a 0.25-m^2 SCOR net (Scientific Committee for Oceanographic Research) that was approximately 0.5 m in diameter and had a 350-μm mesh. From calibrated tows made with a General Dynamics flowmeter, the volume filtered in one tow was approximately 12 m^3, although this value varied by a factor of 2 among hauls. The net was step-towed for about 2 min at each tow depth of 10 m and 5 m and near the surface (6 min total). Most of the basin could be sampled with this protocol; had the maximum towing depth been greater than 10 m, only a small area in the center of the basin could have been sampled by a standard procedure. Samples were fixed in 10% seawater formalin and taken to the Pacific Biological Station (PBS). There, larvae were counted and measured microscopically (snout to end of notochord) to the nearest 0.1 mm. As many as 20 larvae from each sample were rinsed individually, dried at 70°C for 24 h, and weighed to the nearest 0.001 mg.

Tows for larval Pacific herring were made each year, 1982–1984, even though no eggs were transplanted in 1983. This allowed detection of larvae that might have been naturally spawned at undetected sites within the basin or advected from outside it.

Larval control tests.—In 1982, some windrow eggs from the two first wind-drift sources, Deep Bay and Rathtrevor Beach, were taken live to the laboratory (PBS) to ensure that the eggs remained viable and that development progressed normally. The eggs were incubated for the remainder of their developmental period in flowing seawater tanks (10°C, 28‰). After hatching, larvae from these wind-drifted eggs were maintained in a seawater tank (10°C, 28‰). The larvae were not fed. Each day, a sample of 20 larvae was taken from each tank. Each larva was measured to the nearest 0.1 mm, dried, and weighed to the nearest 0.001 mg.

Results

Egg Viability Tests

Eggs incubated out of water on a Vexar screen in the Vancouver Public Aquarium survived better if they received a fine spray than if they were kept damp, but received no spray (Table 1). Similar results were obtained when sprayed and dry eggs were incubated in wooden crates on the beach (Table 2). The unsprayed eggs became much more compacted, their temperature rose, and their survival was zero after 3 d. Survival was high in the sprayed eggs for 4–5 d but then decreased in the crates with greater egg depths. By April 6, the surviving eggs were ready to hatch, but hatch did not occur in the crates. However, when the test was concluded on April 7, the eggs were dumped into the adjacent water and larvae from the released eggs were observed within 1 h.

TABLE 1.—Survival of wind-drifted Pacific herring eggs during incubation tests at the Vancouver Public Aquarium. Groups of 10,000 eggs were incubated out of water. One group was sprayed with seawater, the other was covered with a damp paper towel. Air temperature varied from 15 to 18°C.

Time from start (h)	Sprayed eggs		Damp eggs, no spray	
	N	Survival	*N*	Survival
16	1,143	91.3%	1,017	76.1%
45	1,021	82.9%	~1,000	0.5%
64		~4%[a]		0.0%[b]

[a]About 400 viable larvae hatched from the original 10,000 eggs.
[b]All eggs were disintegrating.

Weight and Numbers of Transplanted Eggs

The estimated weights of eggs transplanted in 1982 and 1984 could be in error by at least 25%. In 1982, 16,580 kg of eggs and material were moved in five shipments (Table 3). The weight of eggs alone was estimated at abcut 9,400 kg, of which about 4,559 kg were estimated to be viable 1–2 d prior to hatching. This last weight of eggs would require a biomass of spawning adults of about 19.15 tonnes, and would result in about 19.15×10^8 larvae. In 1984, about 9,000 kg of material were transported, and the net egg weight was about 5,060 kg. From this we estimated that about 2,782 kg of eggs were viable, which corresponded to a spawning stock biomass of about 11.7 tonnes and an estimated production of 11.7×10^8 larvae. This quantity of eggs is relatively small compared to those from natural spawns.

Numbers and Densities of Larvae

Total numbers of larvae in 1982 and 1984 (Table 4) were estimated as the product of the mean density (number/m^2) by the total area of Hidden Basin (850,000 m^2). A rough estimate of survival is the percentage of the estimated total number of larvae relative to the estimated total number of viable eggs. Maximum survival, estimated at 10.53% on March 26, 1982, represents yolk-sac larvae from the first transplant made on March 15. Eggs from the same batch hatched in the laboratory around March 23, so the larvae caught on March 26 would have been free-swimming for several days at most, and perhaps hatching was not complete in the slightly cooler water (6–8°C) of Hidden Basin. The estimate of 8.94% survival on April 4 is more revealing because it is based mainly on eggs from the transplants of March 15. Few, if any, of the eggs transplanted on March 17 hatched, and April 4 was too soon for the transplants of March 27–29 to have hatched. Therefore, most of the larvae captured on April 4 probably were derived from the March 15 transplant that had hatched around March 23. Daily survival, based on a posthatch period of 12–14 d, was 81.7–84.1%/d (instantaneous mortality, 0.172–

TABLE 2.—French Creek incubation tests of wind-drifted Pacific herring eggs. Eggs were held in six wooden 1-m^3 crates, out of water. Three depths (0.3, 0.6 and 1 m) of eggs were used for two treatments: three crates were sprayed and three not sprayed with seawater. Samples of 500–1,000 eggs were taken at a depth of 15 cm.

	Dry eggs				Sprayed eggs			
		% viability for egg depths of				% viability for egg depths of		
Date	Temperature (°C)	0.3 m	0.6 m	1.0 m	Temperature (°C)	0.3 m	0.6 m	1.0 m
Apr 1		88	88	88		88	88	88
Apr 2	10.5 –12	94	95		7 –8		91	94
Apr 3	14 –16	0	0	0	7 –8	87	89	95
Apr 4					6	91	97	94
Apr 4							85	27
Apr 5						90		50
Apr 5								0
Apr 6						10		
Apr 7							10	

0.201). These estimates, however, do not take into account that some larvae dispersed, probably with tidal movements, from Hidden Basin into the adjacent water of Malaspina Strait.

Maximum survival estimates in 1984 were lower: a total estimated survival of about 0.70% on April 15 corresponds to a daily survival rate of about 71.9% (instantaneous mortality, 0.330) over a 15-d period from April 1, the approximate expected hatching date of the earliest transplants. However, the estimated survival rate based on later larval catches would be much lower.

In all years, some Pacific herring larvae were found outside Hidden Basin, even in 1983 when no transplants were made (Table 5). Nine larvae were found inside Hidden Basin in 1983. For all areas, however, the larval density was much lower in 1983 (Table 5) than it was in the transplant years (Table 4). Also the nine larvae found in 1983 were relatively large, which indicates that they were old enough to have originated elsewhere and emigrated to the Hidden Basin vicinity, which is at least 20 km from the nearest spawning locations (Figure 1) used from 1982–1984. The density of larvae outside of Hidden Basin in 1983 (Table 5) was lower than in 1982 but higher than in 1984. All of the larvae found outside the basin in 1983 were larger and, presumably, older than yolk-sac larvae. Similarly, most larvae found outside the basin in 1982 and 1984 were without yolk sacs and larger than those within the basin, which suggests that most of the youngest, smallest larvae were retained for at least a short time within the basin. However, the presence of nontransplanted larvae outside the basin in 1983 is

TABLE 3.—Source locations, dates, weights, viabilities, and estimated numbers of viable wind-drifted Pacific herring eggs transplanted in 1982 and 1984. The weight of spawning adults (1:1 sex ratio) required to produce the same biomass of transplanted eggs is shown in the last column. The methods used for the early transplants of March 15 and 17 were inappropriate and account for the low and negligible estimate of viable eggs resulting from those transplants.

				% viable eggs			Viable eggs			
Transplant date (source)	Total weight of windrow (kg)	% eggs	Weight of eggs (kg)	Departure	Arrival	Last sample date	%	Weight (kg)	Number (millions)	Weight of adults (tonnes)
1982										
Mar 15 (Deep Bay)	3,600	95	3,420	92	40–60	Mar 31	5	171	72.0	0.72
Mar 17 (Rathtrevor Beach)	680	40	272	61	25	Mar 31	Nil			
Mar 27 (Columbia Beach)	4,300	45	1,935	87	85	Apr 3	80	1,548	650.0	6.50
Mar 29 (Columbia Beach)	4,000	45	1,800	80	80	Apr 3	80	1,440	605.0	6.05
Apr 4 (French Creek)	4,000	50	2,000		70[a]	Apr 9	70	1,400	588.0	
Total	16,580		9,427					4,559	1,915.0	19.15
1984										
Mar 25 (Hornby Island and Deep Bay)	7,600	50	3,800	57	~50		50[a]	1,900	798.0	
Apr 2 (French Creek)	1,400	90	1,260	79	~70		70[a]	882	371.0	
Total	9,000		5,060					2,782	1,169.0	11.69

[a]Presumed value; not estimated directly.

TABLE 4.—Numbers of Pacific herring larvae collected in Hidden Basin, 1982 and 1984. The volume of water filtered with each tow was estimated at 12 m^3. The oblique tows were limited to the upper 10 m of water so the estimated larval densities per unit area are about 10 times those per unit volume. Total area of the basin was calculated to be about 850,000 m^2. Estimates of larval survival are based on the percentages of estimated larvae relative to the estimated number of viable and hatched eggs present in Hidden Basin on the date of the tow.

Date of capture	Number of		Number of larvae per			Number of larvae (millions)		Apparent survival (%)
	Tows	Larvae	Tow	m^3	m^2	Total	Hatched	
1982								
Mar 25	8	9	1.1	0.091	0.91	0.773	72.0	1.07
26	4	43	10.7	0.892	8.92	7.582	72.0	10.53
Apr 4	4	670	167.5	13.958	139.58	118.640	1,327.0	8.94
5	4	19	4.7	0.392	3.92	3.332	1,327.0	0.25
7	5	449	89.8	7.483	74.83	63.605	1,915.0	3.32
10	4	108	27.0	2.250	22.50	19.125	1,915.0	1.00
14	3	12	4.0	0.333	3.33	2.830	1,915.0	0.15
21	4	44	11.0	0.962	9.62	8.177	1,915.0	0.43
1984								
Apr 5	5	7	1.4	0.117	1.17	0.994	1,169.0	0.08
11	5	3	0.6	0.050	0.50	0.425	1,169.0	0.04
15	5	58	11.6	0.967	9.67	8.220	1,169.0	0.70
16	5	6	1.2	0.100	1.00	0.850	1,169.0	0.08
20	5	0	0.0	0.000				
23	5	2	0.4	0.030	0.30	0.225	1,169.0	0.02

evidence that some larval Pacific herring can reach the vicinity of Hidden Basin from other spawning locations.

Larval Growth and Feeding

The expected mean dry weights and lengths of recently hatched larvae, based on subsamples of wind-drifted eggs incubated in the laboratory, were about 0.12–0.13 mg and 7.8–8.1 mm, respectively (Figure 3). For a 9–11-d posthatching period, without feeding, the laboratory larvae increased their length to a maximum of about 9.3 mm, but their dry weights decreased to about 0.1 mg or less. Daily survival was not monitored during this time but there was no substantial mortality (i.e., it was not over 10%) until the termination of the test in early April, when most larvae died within a few days. We cannot explain the small differences in mean weights between the two groups of larvae maintained in the laboratory. However, these differences in the lengths and weights of the two groups of laboratory-reared larvae are slight when contrasted to the lengths and weights of larvae from Hidden Basin (Figure 4). Most of the field-captured larvae had lengths and weights that exceeded those of the laboratory specimens, which indicates that they had been feeding. We cannot trace a single larval cohort, corresponding to a specific transplant, from mean lengths and weights because of the overlap of larval sizes among transplants. However, frequency distributions of dry weights indicate sev-

TABLE 5.—Summary of catches and densities of Pacific herring larvae outside Hidden Basin in 1982–1984, and inside Hidden Basin in 1983. Numbers in parentheses are standard errors.

Date and year	Number of			Mean length (mm)	Mean weight (mg)
	Tows	Larvae	Larvae/m^2		
Apr 14, 1982	7	354	6.321	13.43 (2.51)	0.474 (0.027)
Apr 26, 1982	6	18	0.371		
Apr 15, 1983 In Hidden Basin	3	9	0.042	14.39 (10.90)	0.459 (0.111)
Outside Hidden Basin	14	142	1.260	14.41 (2.40)	0.724 (0.052)
Apr 5, 1984	4	24	0.750	12.69 (2.73)	0.442 (0.028)

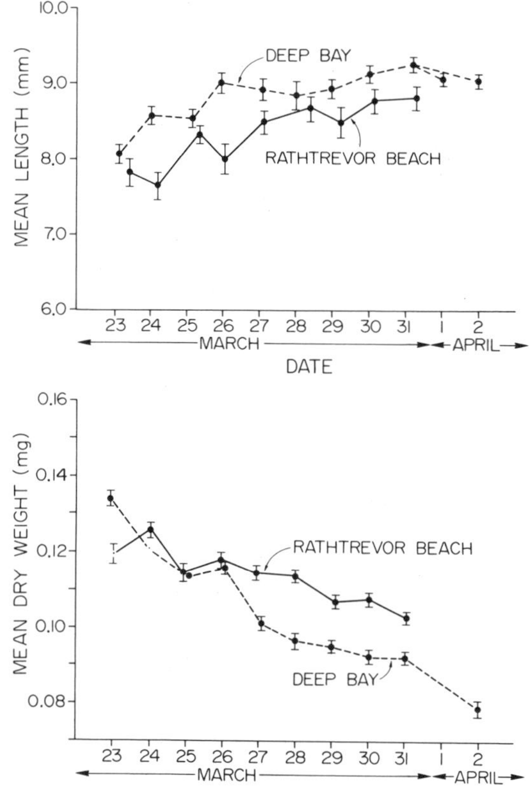

FIGURE 3.—Mean lengths and dry weights of larval Pacific herring reared in the laboratory (without food) from wind-drifted eggs taken from Deep Bay and Rathtrevor beaches in 1982. The vertical lines indicate the standard errors about the mean for sample sizes of 20.

eral larval cohorts that can be followed over time, although the distinction between successive modes is not sharp (Figure 5). Most of the field-collected larvae weighed more than 125 μg, indicative of feeding, although some, especially those captured on April 7 and 10, had lost weight, and a few weighing less than 75 μg probably were starving.

History of Herring Spawning in the Transplant Area

Records made by fishery officers show that Pacific herring spawned naturally in the vicinity of Hidden Basin for two periods: 1948–1951 and 1960–1972 (Figure 6). If it is assumed that, on average, there were about 500,000 eggs/m^2 (a density frequently observed), the maximum spawning biomass required for the observed deposition would have been only 200 tonnes in the highest years of 1960 and 1971 and much less in other years. We conclude that spawning in Hidden Basin and the adjacent area of Blind Bay, where most of the spawn deposition occurred, was intermittent over the years and relatively small (compared to other areas) in the years when spawning occurred there. Spawn deposition by Pacific herring is not geographically random; indeed the repetitive spawnings for the 13 years from 1960 to 1972 occurred within the same 5 km length of shore, and there was no other spawning for 50 km in either direction. During those years, therefore, there must have been some factor or factors that caused Pacific herring to spawn in Blind Bay and Hidden Basin. If those same factors are attractive to Pacific herring and could operate on fish derived from transplanted eggs, then transplants might have the potential to generate a repetitively spawning stock.

Discussion

Egg and Larval Survival

There is little question that transplanted Pacific herring eggs successfully incubated and hatched within Hidden Basin. There has been no documented spawning by this species in Hidden Basin or the immediate vicinity since 1972, and the few larvae caught in 1983 (a nontransplant year) were so large that they undoubtedly originated outside the basin.

Comparisons of lengths and dry weights of captured and laboratory-reared larvae indicate that the transplanted larvae in Hidden Basin were feeding and growing. Their abundance, however, decreased over time. The several estimates of mean daily survival (71–84%/d within the basin) and of instantaneous mortality (0.17–0.30) are within the ranges previously estimated for Pacific herring and summarized by McGurk (1986). In a study of larval Atlantic herring *Clupea harengus harengus* that were similar in age to our Pacific herring, Henderson et al. (1984) reported instantaneous mortalities of 0.061–0.074, which are much lower than those of our transplanted larvae. However, our mortality estimates could be inflated if larvae dispersed from Hidden Basin to outside waters.

Transplant Technology

Wind-drifted spawn as a source of eggs for transplantation has the advantage of being free, that is, most such eggs would not survive to hatching unless they were transplanted. Therefore, these eggs can be used without reducing the

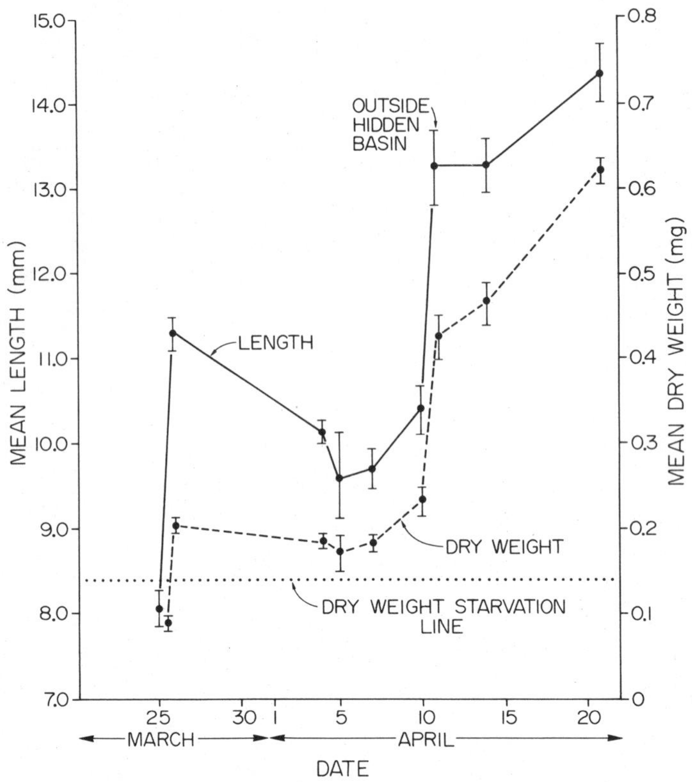

FIGURE 4.—Mean lengths and dry weights for Pacific herring larvae captured in 1982 in Hidden Basin, March–April 1982. The vertical lines indicate the standard errors about the means for sample sizes of 20 or fewer. Dry weights below the horizontal dotted line indicate starvation or substantial weight loss compared to hatching weights. All catches on April 11 and most of those on April 14 were made outside Hidden Basin.

potential for recruitment that would occur if normal spawn had been used. If wind-drifted spawn is used, however, it must be checked to ensure that a sufficient number of eggs are viable. Also, handling during transfer will affect the survival of eggs. Our results indicate that immersion of eggs in nonflowing seawater results in high mortality, probably because of the high biological oxygen demand of the eggs and associated organic beach debris, which often consists of dead or dying macrophytes (Hay and Miller 1982). Continuously soaking piles of wind-drifted eggs with a seawater hose can be deleterious because necrotic debris may be compacted among eggs. The least mortality occurred when eggs were maintained in air but kept cool and moist with an intermittent, light seawater spray. These eggs continued to incubate under conditions similar to those of naturally incubating eggs in the intertidal zone. This method would not necessarily apply to eggs spawned on submerged artificial substrates, such as that used by Moser and Hansen (1986). They used naturally spawned eggs on a moveable substrate that could be relocated after spawning was completed. However, the objectives of their methods and ours are the same.

Biological and Economic Value of Transplants

At the present time we are uncertain about the number and geographic distribution of Pacific herring stocks in the Strait of Georgia. If the Strait of Georgia maintains only one large migratory stock, it is hard to justify transplants of wind-drifted eggs because potential benefits would not warrant the costs. For instance, in 1982 the spawning equivalent of about 19 tonnes of fish was moved. Presumably, if that spawn had not been transplanted, most of it would have perished, so the effect of the transplant was to add the spawning equivalent of about 19 tonnes of ripe adults to the population. The same effect would

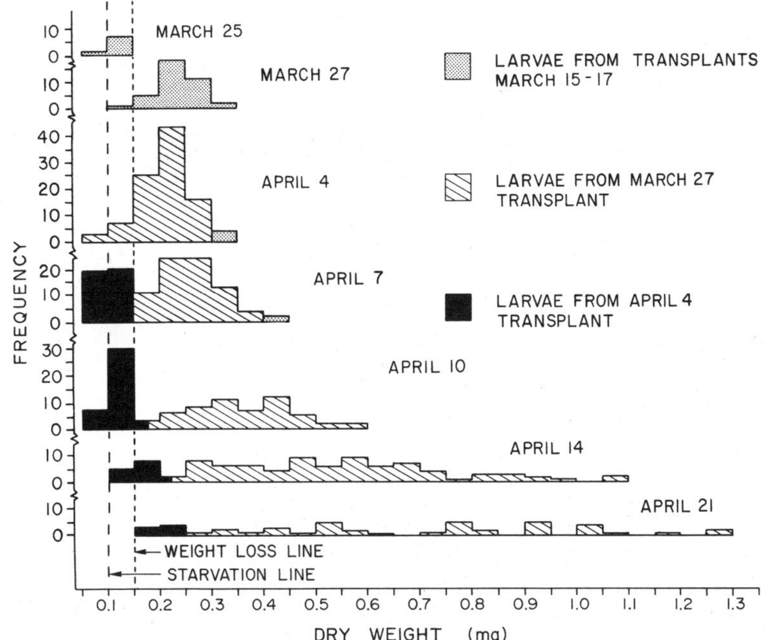

FIGURE 5.—Frequency distributions, by date, of dry weights of larval Pacific herring collected in Hidden Basin in 1982. Several larval cohorts are indicated by different shading, but there is some overlap between cohorts that cannot be differentiated. The vertical broken lines mark dry weights below which weight loss or starvation are likely to occur.

have been realized if the wind-drifted spawn had been ignored, and the fishery had been reduced by 19 tonnes to allow spawning of some fish that would otherwise have been captured. In practice, management of catch rates to this level of precision would have been impossible: the 1982 Strait of Georgia fishery took over 9,000 tonnes (by quota) from a total biomass estimated at more than 50,000 tonnes (Anonymous 1982). Nevertheless, for the purpose of illustration, the cost of a 19-tonne reduction, based on the average value of Pacific herring of Can$1,000–2,000/tonne (US$750–1,500/tonne) (Anonymous 1983) would be about Can$20,000–$40,000. This return is approximately equal to the cost of the transplant if all costs, direct and indirect, are considered. In addition, if the Strait of Georgia supports only one large stock, the same benefit would be achieved by simply returning eggs to the water in the immediate vicinity of the source or of the wind drifts. A policy of returning wind-drifted eggs to the water presumes, however, that it is both biologically and economically worthwhile to add more eggs to the Strait of Georgia or to any spawning stock of fish. The assumption is that adding or salvaging eggs will result in increased recruitment or cohort strength in future years. This assumption may not be valid (except at very low biomass levels) because strong year classes apparently can be derived from years with relatively low spawn deposition, and vice versa (Taylor 1964).

The biological basis for repetitive use of a spawning ground by Pacific herring is uncertain. It may involve either some aspect of the physical environment yet to be identified or a complex combination of physical and biological factors such as temperature, salinity, macrophyte composition,

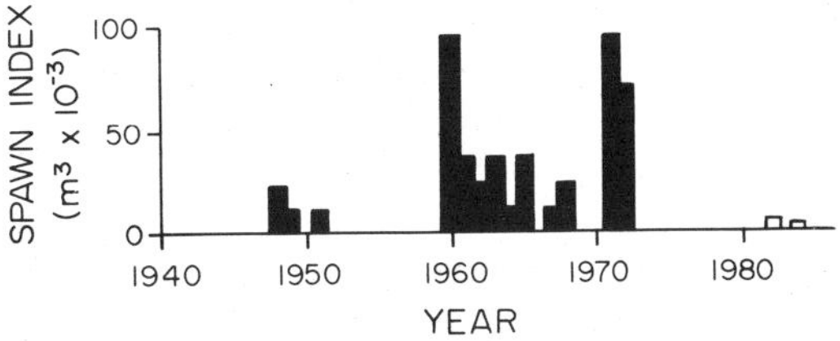

FIGURE 6.—Spawning history of Hidden Basin and Blind Bay. The bars indicate the magnitude of the spawn index, a measure of relative spawn deposition. The open bars for 1982 and 1984 indicate the approximate indexes that would have resulted if the transplanted spawn in those years had been naturally deposited.

substrate, and current (Haegele and Schweigert 1985; Hay 1985). Clearly, Pacific herring are not like salmon and some other anadromous species in which olfactory imprinting of juveniles results in precise recognition of natal streams (see, for example, Harden Jones 1968). Probably the same homing precision is not possible for species spawning in marine coastal areas because tidal mixing of surface waters would disrupt any unique organic bouquet that shoreline areas might have.

The biological benefits and cost effectiveness of transplants would be clearer if we could be certain that larvae derived from wind-drifted spawn remained resident in the vicinity of the transplant location. If small, local stocks already exist alongside large, migratory stocks within the Strait of Georgia, these would offer encouragement that transplants could be successful because the transplant might establish a stock where one previously did not exist. This would be desirable because total Pacific herring biomass would be increased, thereby diminishing the effect of the fisheries and providing more prey for commercially and recreationally important species such as salmon. The cost effectiveness of transplants would be increased because the transplant costs could be amortized over the time that the new stock remains viable.

The evidence for the existence of small local stocks, however, is equivocal. Analysis of spatial and temporal spawning patterns indicates that there could be as many as 40 or more stocks along the British Columbia coast (Hay 1986) and perhaps several small stocks within the Strait of Georgia, in addition to one or two larger stocks. Currently, stock assessments for the commercial fishery are made separately for southern and northern spawning areas within the strait (Haist et al. 1986). This division is based on a variety of factors such as analyses of morphology (Meng and Stocker 1984), gill-net selectivities (Hay et al. 1986), and spawning times (Hay 1986). The evidence for the existence of smaller stocks in southern Georgia Strait is less certain, and it would be unwarranted to conclude that transplants could, or could not, establish new stocks or reestablish lost stocks.

Criteria of Transplant Success

The criteria for success of transplants are related to the arguments for cost effectiveness and concepts of the Pacific herring stock structure in the Strait of Georgia or wherever Pacific herring transplants are made. If only one large stock exists in Georgia Strait, one might argue that production of viable feeding larvae is an adequate criterion of success. This alone, however, is not sufficient justification for transplants because the same result would be achieved by returning eggs to the water in the vicinity of the wind drifts. If the stock structure is rich, with many local stocks in Georgia Strait, the only definitive criterion of success is the establishment of a new, resident spawning group of Pacific herring in a transplant location where the species has not previously existed or has been absent for years. In this project, no Pacific herring were observed spawning in Hidden Basin either in 1985, as newly recruited 3-year-olds derived from the 1982 transplant, or in 1987, as recruits from the 1984 transplant.

Future of Herring Spawn Transplantation: Benefits and Risks

There still is an enormous interest in the potential for Pacific herring spawn transplants, mainly by sport-fishery groups who feel that enhancing local Pacific herring abundance will benefit local sport-fishing operations. To this end, the Department of Fisheries and Oceans has supported future attempts to move Pacific herring spawn to Indian Arm Inlet, north of Vancouver, Canada, where Pacific herring were known to have spawned in past years but not in recent times. This inlet also is geographically remote from the nearest documented spawning areas. Therefore, if transplants of spawn to Indian Arm or elsewhere lead to future spawning, it would be very strong evidence that transplants can be used to establish new stocks.

Finally, it should be appreciated that there is a risk associated with continued transplant tests. The risk involves creating the mistaken impression among the public, politicians, resource managers, and industrial developers that this approach can be used to mitigate the effects of industrial or recreational developments on established spawning areas. Proponents of such developments could argue that the Pacific herring spawn transplant approach could be used to comply with the Canada Fisheries Act (or similar legislation in the USA or other nations), which states that there must be no net loss of fish habitat. The methods of transplanting wind-drifted spawn of Pacific herring still are primitive and unproven. The approach cannot be used or cited, at the present time, as a method to compensate for destruction of established spawning habitat.

Acknowledgments

The funding for this project was primarily via contracts from the Resource Development Branch, Department of Fisheries and Oceans, to Tidal Rush Marine Farms of Nelson Island. Brad Hope, Ken Hope, and June Hope participated in the collection and transport of Pacific Herring eggs and in larval surveys. Don McRae and Addie McRae assisted with aerial surveys to locate wind-drifted eggs, the collection and transport of eggs, and the egg-incubation tests at the French Creek Marina. Bruce McCarter assisted with laboratory analysis of larval samples at the Pacific Biological Station.

References

Anonymous. 1982. Review of the 1981–82 British Columbia herring fishery and spawn abundance. Canada Department of Fisheries and Oceans, Information Bulletin, Vancouver.

Anonymous. 1983. Review of the 1982–83 British Columbia herring fishery and spawn abundance. Canada Department of Fisheries and Oceans, Information Bulletin, Vancouver.

Haegele, C. W., R. D. Humphreys, and A. S. Hourston. 1981. Distribution of eggs by depth and vegetation type in Pacific herring (*Clupea harengus pallasi*) spawnings in southern British Columbia. Canadian Journal of Fisheries and Aquatic Sciences 38:381–386.

Haegele, C. W., and J. F. Schweigert. 1985. Distribution and characteristics of herring spawning grounds and description of spawning behaviour. Canadian Journal of Fisheries and Aquatic Sciences 42 (Supplement 1):39–55.

Haist, V., J. F. Schweigert, and M. Stocker. 1986. Stock assessments for British Columbia herring in 1985 and forecasts of the potential catch in 1986. Canadian Manuscript Report of Fisheries and Aquatic Sciences 1889.

Harden Jones, F. R. 1968. Fish migration. Arnold, London.

Hay, D. E. 1985. Reproductive biology of Pacific herring (*Clupea harengus pallasi*). Canadian Journal of Fisheries and Aquatic Sciences 42 (Supplement 1): 111–126.

Hay, D. E. 1986. A stock hypothesis based on spawn and winter distribution. Canadian Manuscript Report of Fisheries and Aquatic Sciences 1871:145–148.

Hay, D. E., K. D. Cooke, and C. V. Gissing. 1986. Experimental studies with Pacific herring gillnets. Fisheries Research (Amsterdam) 4:191–211.

Hay, D. E., and A. R. Kronlund. 1987. Factors affecting the distribution, abundance, and measurement of Pacific herring (*Clupea harengus pallasi*) spawn. Canadian Journal of Fisheries and Aquatic Sciences 44:1184–1194.

Hay, D. E., and D. C. Miller. 1982. A quantitative assessment of herring spawn lost by storm action in French Creek, 1980. Canadian Manuscript Report of Fisheries and Aquatic Sciences 1636.

Henderson, P. A., J. W. Whitehouse, and G. H. Cartwright. 1984. The growth and mortality of larval herring, *Clupea harengus* L., in the River Blackwater estuary, 1978–1980. Journal of Fish Biology 24:613–622.

Hourston, A. S. 1982. Homing by Canada's west coast herring to management units and divisions as indicated by tag recoveries. Canadian Journal of Fisheries and Aquatic Sciences 39:1414–1422.

Iles, T. D., and M. Sinclair. 1982. Atlantic herring: stock discreteness and abundance. Science (Washington, D.C.) 215:627–633.

McGurk, M. D. 1986. Natural mortality of marine pelagic fish eggs and larvae: role of spatial patchiness. Marine Ecology Progress Series 34:227–242.

Meng, J. M., and M. Stocker. 1984. An evaluation of morphometrics and meristics for stock separation of Pacific herring (*Clupea harengus pallasi*). Canadian Journal of Fisheries and Aquatic Sciences 41: 414–422.

Moser, M., and J. E. Hansen. 1986. An artificial substrate and holding cages for herring spawn. North American Journal of Fisheries Management 6:603–604.

Taylor, F. C. H. 1964. Life history and present status of British Columbia herring stocks. Fisheries Research Board of Canada Bulletin 143.

American Fisheries Society Symposium 5:60–67, 1988

Estimation of Hatch Periods for Yellow Perch, Based on Otolith Readings from Juveniles (Age-0)[1]

PERCE M. POWLES

Biology Department, Trent University, Peterborough, Ontario K9J 7B8, Canada

STANLEY M. WARLEN

National Marine Fisheries Service, Southeast Fisheries Center
Beaufort Laboratory, Beaufort, North Carolina 28516, USA

Abstract.—The daily rings on otoliths of yellow perch *Perca flavescens* provided a reliable means for back-calculating growth and estimating hatching dates. The first microzone was formed 1–3 d posthatch in larvae reared in the laboratory for 12 d posthatch. Subsequently rings formed daily, but in otoliths of yolk-sac larvae, daily rings were often accompanied by subdaily rings. The aging technique for yellow perch larvae was validated from a field sample of larvae ranging in age from 29 to 35 d taken 1 month after their known hatch date. Yellow perch egg strands incubated in situ (at a depth of <1 m) showed slightly longer developmental times (140 versus 120 degree-days at 15–20°C) than constant temperature laboratory incubation studies reported elsewhere. The duration of estimated hatch periods for yellow perch eggs in four lakes ranged from 6 d in the most southern lake to over 2 weeks in the most northern lake.

In the extensive literature on yellow perch *Perca flavescens*, including a synopsis of biological data (Thorpe 1977), no one has yet reported yellow perch hatch or spawning times deduced from daily otolith ring records. Daily otolith aging has not been validated for otoliths from young-of-the-year fish. A unique feature of yellow perch reproduction is the egg strand, which is usually strewn within the littoral zone in a contracted period of time, and is easily observed by snorkeling. Although in situ incubation experiments most closely simulate nature, only one in situ incubation rate experiment for yellow perch has been documented (Mansueti 1964). Others (Hokanson and Kleiner 1974) have incubated yellow perch eggs at various constant temperatures in the laboratory.

Although there have been a number of studies (Warlen 1982; Brothers et al. 1983; Miller and Storck 1984; Graham and Orth 1986) that have used daily otolith growth rings to estimate spawning times, the technique has not been used for yellow perch. This study reports on the application of daily otolith growth increment counts to estimate time of spawning and hatch periods. Such information on spawning dates along with information on conditions for incubation may be important in determining success of year classes of yellow perch.

[1]Reference to trade names does not imply endorsement by the National Marine Fisheries Service.

The objectives of this study were to (1) assess the reliability of daily otolith rings as indicators of the age of juvenile yellow perch, (2) estimate in situ (in-lake) incubation rates of yellow perch eggs for predictive purposes, (3) determine the age at first daily ring or microzone formation on yellow perch otoliths and, (4) estimate hatch duration, spawning dates, and subsequent growth of larval and early juvenile yellow perch.

Methods

Field sampling.—Young perch were collected in 1983 from four Ontario lakes. The methods of capture, dates, and locations of all samples are shown in Table 1. The lakes were chosen to represent a south to north gradient. By a priori reasoning, we constructed two null hypotheses: (1) a difference in sequence of spawning would not occur among yellow perch, and (2) fish from the warmest littoral zones would not spawn first.—Our most southerly sample was from Chemung Lake (mean depth, 3 m), a warm, shallow, eutrophic limestone lake near Peterborough. Jack Lake (mean depth, 5.4 m) is a mesoeutrophic, slightly cooler lake 50 km north of Peterborough. Clear and Crown Lakes are acid-sensitive oligotrophic lakes (mean depth, 12.5 m and 8.0 m, respectively) 120 km north of Peterborough, with spring temperatures an average of 5°C cooler than Chemung Lake in the littoral zone in early May.

TABLE 1.—Location, time, and gear for collections of yellow perch from four Ontario lakes.

Lake	Date	Standard length range (mm)	Gear	Number captured	Number of fish aged
Chemung	6 Jun 1983	14.8–16.1	Light trap	400	25
Jack	24 Jun 1983	19.4–23.3	Beach seine	1,000	25
Clear	24 Jul 1983	26.0–41.0	Beach seine	200	25
Crown	21 Jul 1983	21.5–32.0	Beach seine	86	25
Clear	8 Jul 1984	23.0–33.0	Beach seine	60	19

Age validation.—In 1982, light-trap catches showed that yellow perch hatched in Chemung Lake the week of May 3 at 18°C (Gregory and Powles 1985). In 1983, larvae were hatched during the week of May 6 at 17°C ambient water temperature. Using the methods of Gregory and Powles (1985), we collected some of these larvae 1 month later on June 6, 1983 (Table 1). These 1-month-old larvae served as an in situ control group of wild fish of known age that we used to validate daily increment deposition in otoliths, a validation method listed by Geffen (1987). All yellow perch collected were preserved in 70% ethanol, and the standard length was measured to the nearest 0.1 mm.

We also compared sizes observed over time from light-trap catches (Gregory 1983) with sizes back-calculated from daily otolith rings. Thus, the sizes compared were from 1982 light-trap catches and 1983 back-calculated lengths-at-age (in days).

We observed perch egg strands in Clear Lake on May 26–27, 1984, at 10–12°C (to 3 m depth). We do not know the exact date of spawning nor when hatch or emergence took place, but this gave us a benchmark date to help assess the accuracy of our back-calculations of hatch times.

To estimate incubation times, we used the average of the developmental rates found by Hokanson and Kleiner (1974), which were 128 degree-days (6.5 d at 19.7°C) and 120 degree-days (8 d at 15°C), and our own in situ rate of 140 degree-days.

To estimate the age of larvae at first otolith growth increment formation (Fives et al. 1986), we reared eggs under three temperature regimes, 22, 16, and 10°C. Eggs were obtained from several strands in Chemung Lake at 1 m depth (16°C) on April 18, 1986. Larval samples from each group were collected daily until 1 week after hatching. Observed times of swimming activity and exogenous feeding after hatch were recorded. The photoperiod was maintained at 12 h dark:12 h light.

Aging.—Otoliths (sagittae) from smaller larvae (<15 mm standard length) were removed, cleaned in distilled water, and mounted on a microscope slide with Flo-Texx mounting medium. Otoliths from larger larvae were removed, placed convex side up on a very small drop of Krazy Glue. The convex side was then ground with number 600 grit wet sandpaper glued to another glass slide. Care was taken not to overgrind the otolith and thus eradicate the nucleus. When the core and inner rings were visible at 100× magnification, the otolith was polished with diamond paste, and then washed.

Otoliths were examined under oil with a compound microscope fitted with a television camera (600× or 1,000×). Two persons made independent counts from images of otoliths on a video

TABLE 2.—Ambient water temperatures, cumulative degree-days,[a] dates, and hatch of yellow perch, eggs held in situ in Chemung Lake, 1987 (water depth, 25 cm).

Variable	April dates								
	15	16	17	18	19	20	21	22	23
Temperature									
Maximum (°C)	15.4	15.8	16.1	16.7	21.1	23.3	22.8	20.1	18.1
Minimum (°C)		15.0	14.4	13.9	15.0	21.6	22.0	19.4	16.5
Mean (°C)	15.4	15.4	15.3	15.3	18.1	22.5	22.4	19.8	17.0
Cumulative degree-days	15.4	30.8	46.1	61.4	79.5	102.0	124.4[b]	144.2	
Observations			Movement apparent		Eyed	Partial hatch	50% hatch		Total hatch

[a]Sum of mean daily temperatures above 0°C.
[b]Some dead eggs in replicate.

TABLE 3.—Mean number of rings observed on otoliths of embryonic and hatched yellow perch maintained at three temperature regimes. Eggs were collected from several strands in Chemung Lake, Ontario, on April 18, 1986 (temperature 16°C).

	Warm (20–22°C)			Medium (15–16°C)			Cold (10°C)		
Days posthatch	Date (Apr–May)	Mean number of rings	Number of fish aged	Date (Apr–May)	Mean number of rings	Number of fish aged	Date (Apr–May)	Mean number of rings	Number of fish aged
				Embryonic or prehatch period					
							22	0.0	5
							23	0.0	5
				22	0.0	5	24	0.0	5
				23	0.0	5	25	0.0	5
				24	0.0	4	26	0.0	5
				25	0.0	5	27	0.0	5
	18	0.0	5	26	0.0	5	28	0.0	4
	19	0.0	5	27	0.0	4	29	0.0	5
				Hatching period					
	20	0.0	5	28	0.5	3	30	0.0	5
	21	0.5	5	29	0.5	4	1	0.0	5
				Posthatch period					
1	22[a]	0.6	5	30	0.7	4	2	0.5	4
2	23	1.0	3	1[a]	1.8	5	3[a]	1.8	5
3	24	1.8	4	2	3.0	5			
4	25[b]	2.3	4	3	4.4	5			
5	26	4.5	4						
6	27	5.0	4						
7	28	6.2	6						
8	29	6.8	6						
9	30	7.4	5						
10	1	9.4	5						
11	2	10.0	5						
12	3	10.9	5						

[a]Swimming noted on these dates.
[b]Exogenous feeding noted on this date.

monitor. In cases of disagreement, the average age was taken. An electronic cursor was used to measure the distance between growth rings, and a growth curve was determined for each fish.

The relationship between otolith dimension and fish length was determined empirically before making the back-calculations for growth history. Otolith radii were measured (100×) along the long axis (rostral tip to midnucleus) to the nearest 0.01 mm with an ocular micrometer.

Incubation in situ experiment.—There are only a few studies of incubation rates that allow us to compute approximate hatch time (Raney 1959; Mansueti 1964; Hokanson and Kleiner 1974; and Guma'a 1978). These range from 30 d at 8°C to 6 d at 20°C. To obtain an estimate of elapsed incubation time in situ for yellow perch, two replicate egg masses of recently spawned eggs (yellow perch spawn at night) were placed in two wire-mesh enclosures in a small bay in Chemung Lake on April 15, 1987. Water depth was 25 cm, and maximum–minimum water temperatures were read daily between 1400 and 1500 hours. Eggs were observed for movement, eye development, and hatch. Samples were preserved in alcohol and formalin every day during April 15–23 to ascertain if eggs were developing normally. Unseasonably high temperatures occurred on April 19, followed by more normal temperatures (Table 2).

Results

Laboratory Reared Eggs

Samples of yellow perch eggs maintained at three constant temperature regimes showed virtually no ring formation until after hatching (Table 3). First growth increment rings were present on at least half the otoliths 1 d after hatch for all temperature regimes. As expected, the 22°C eggs hatched first (April 20–21), the 16°C eggs hatched next (April 28–29), and the cold-incubated eggs hatched last (April 30–May 1). Rings were well spaced and clearest for warm-incubated (22°C) fish and closely spaced and more difficult to distinguish in the group incubated at the coldest (10°C) temperature. The mean number of rings

TABLE 4.—Estimated hatch dates in 1983 from four lakes in Ontario, obtained by back-counting the daily otolith rings in yellow perch. Lakes are arranged from most southerly (Chemung) to most northerly (Clear and Crown).

Lake	Range of number of otolith rings	Estimated hatch periods	Known or observed		Estimated spawning period
			Spawning	Hatch	
Chemung	29–35	May 3–9		May 6	Apr 26–28
Jack[a]	36–45	May 10–19			
Clear	46–64	May 21–Jun 8			May 9–11
Crown[a]	41–56	May 26–Jun 10			
Clear[b]	28–38	May 31–Jun 10	May 26, 27[c]		May 14–16

[a]No ambient temperature records available.
[b]1984 sample.
[c]Eggs seen in littoral zone.

was usually about one less than posthatch age of warm-incubated eggs; the number of rings for medium- and cold-incubated fish corresponded more closely to the posthatch age. However, these series were not very long because of larval mortality. Some swimming was noted on day 1 or 2, depending on temperature.

Samples of Known Age

The first daily ring was assumed empirically to form at hatch. However, daily rings actually commenced from 1 to 3 d posthatch. Our field sample taken 1 month after known hatch further corroborated that rings formed on a daily basis (Table 4) for that period (29–35 days).

Incubation

Our in situ estimate of incubation rate (Table 2) for yellow perch eggs showed that total hatch occurred in 144 degree-days. This is close to Mansueti's (1964) observation of 140 degree-days but higher than the 120–130 degree-days observed by Hokanson and Kleiner (1974) for fish reared under constant conditions.

Estimation of Hatch Periods

Back-calculation of hatch times showed that yellow perch hatched over a range of days and that hatch times varied from north to south (Table 4). Chemung Lake yellow perch were the earliest to hatch, and they hatched over a 6-d period (May 3–9). Jack Lake yellow perch were next (May 10–19), and Clear and Crown lakes fish showed the latest hatch dates (May 21–June 10). The 1984 hatch from Clear Lake occurred over a shorter time period (10 d) than the 1983 hatch (17 d), but it commenced later (May 31 versus May 21). The control or known hatch sample (May 3, 1983) in Chemung Lake, taken by light trap, agreed well with the back-calculated hatch of May 3–9.

We had only one observation date of spawned eggs in situ, that of Clear Lake in 1984. Temperature profiles for Clear Lake (0–5 m) indicated a mean bottom temperature of 12.5°C in mid-May prior to hatch. Incubation should have been about 10 d at this ambient temperature. This observation would place spawning within the May 21 to June 1 period, and indeed eggs were observed May 26–27.

Growth Rates

There was a linear relationship between otolith radius and fish lengths (Figure 1). This relationship allowed us to back-calculate lengths at earlier growth history stages and obtain growth rates for fish from each lake. Among lakes there were no significant differences in relationships of lengths of otoliths to lengths of fish ($P < 0.05$).

Back-calculated lengths and growth rates confirmed that daily rings yield accurate growth estimates but showed some variation between lakes (Figure 2). There were no significant differences in the estimated standard lengths (8–9 mm) of larvae at formation of ring 1. From ring 3 onward, mean length-at-age became different ($P < 0.05$) for early growth (up to age 35 d). Fish from Jack and Crown lakes were not significantly different ($P < 0.05$), but Clear Lake fish maintained the largest mean length-at-age. All yellow perch young grew exponentially, as suggested by Guma'a (1978), and in much the same pattern of increase in length over time as found by Zwanenburg (1981). Our field growth estimates from light-trap samples in 1982 showed early agreement with back-calculated growth for Chemung Lake of 1983 (Figure 2; Gregory 1983).

Qualitative Features

We could not detect daily rings formed before hatch (Figure 3A). Subdaily rings were some-

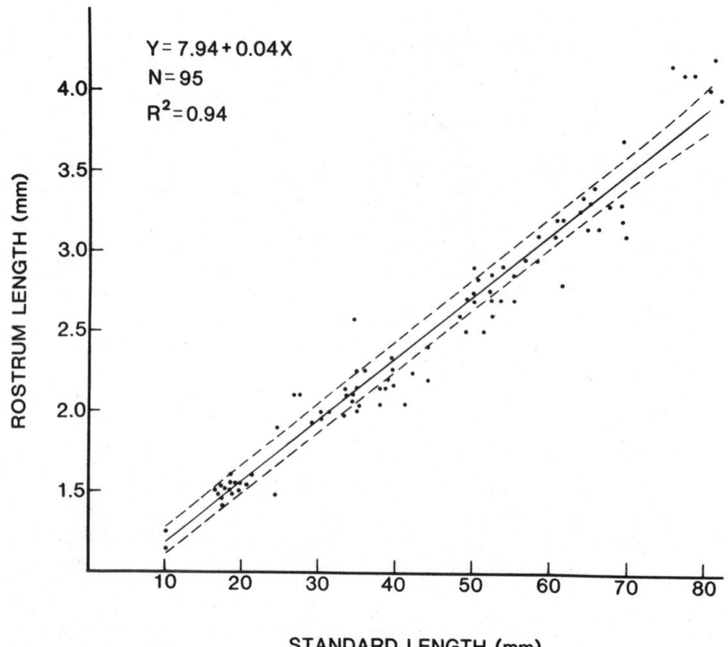

FIGURE 1.—Relationship between otolith (sagitta) length along rostrum and body standard length for yellow perch from central Ontario lakes. Dashed lines are 95% confidence limits.

times seen before and after hatch. A posthatch otolith with two daily rings (Figure 3B) illustrates that the first daily ring was often very distinct. The average distance from the center of the nucleus to this first ring measured 22.1 μm (range 13.8–28.0 μm). Otoliths with fewer than 30–35 rings could be seen by light microscope without grinding (Figure 3C).

Figure 3D shows a typical unground otolith of yellow perch from Chemung Lake. The innermost rings were quite regular and easily seen, but rings toward the margin were further apart and less distinct. Otoliths were clearest if removed and mounted soon after capture (S. Finucan, Trent University, personal communication) rather than after extended storage in alcohol. Areas of crossovers and double nuclei were often observed in otoliths of yellow perch from the two acid lakes. These resembled the peripheral nuclei reported by Campana (1984).

Discussion

Swim-up attempts were noted as early as 1 d posthatch for warm-reared (20°C) yellow perch larvae, and 2 d posthatch for medium- and cold-reared (15 and 10°C, respectively) reared larvae. This swim-up chronology is 4 d earlier than that noted in past studies of *Perca flavescens*. Most other studies reported actual swim-up stage at 5–7 d posthatch (Houde 1969; Hokanson and Kleiner 1974; Ross et al. 1977). While this behavior pattern is almost certainly temperature dependent and associated with adsorption of the yolk, swim-up attempts in yellow perch are almost as significant (in terms of activity) as swim-up stage when the fish are first capable of selecting depths. Gregory (1983) noted that early yolk-sac larvae showed considerable active movement and directed swimming behavior associated with light traps in the field. Conversely, others have reported that yellow perch yolk-sac larvae are inactive (Houde 1969). It may be that such activity (usually observed in daylight in the laboratory) determines the diurnal metabolic cycle that produces daily rings. Early larvae are observed infrequently at night in the laboratory, and they may be fairly quiescent at night. While such early larvae may be swimming up and floating down, they have not yet filled their air sacs. Yellow perch larvae showed considerable activity 1–3 d after hatch, and exogenous feeding occurred 3–5 d posthatch. Similarly, Mansueti (1964) reported first feeding at 3 d posthatch (temperature unspecified).

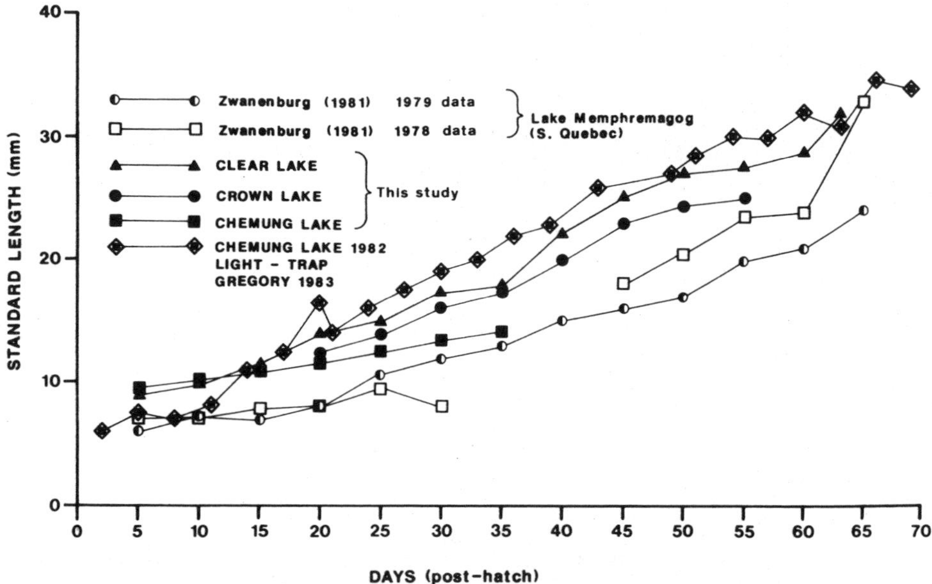

FIGURE 2.—Back-calculated mean length-at-age (in days) for yellow perch from three lakes in Ontario and from Lake Memphremagog in Quebec.

Hatch Period

The hatch period for Chemung Lake was 6 d. This lake was the shallowest and most uniform in depth (1–4 m). Clear Lake had the longest hatch period, 17 d. Of all the lakes sampled, Clear Lake had the steepest littoral zone. If some eggs were strewn and fertilized below the epilimnion, their period of development would be twice as long as those in upper warmer water layers. This occurrence could explain the long 17-d hatch period. After searching thoroughly throughout the lake in May 1984, we found only one egg strand in Clear Lake. That egg strand was found in a 4 m of water at a temperature of 9°C and would have required over 2 weeks to hatch.

There was a latitudinal sequence in the hatching of eggs from south to north in 1983, and the null hypothesis was rejected. The estimated hatch dates were separated by a little over 3 weeks (May 3 for Chemung Lake versus May 26 for Crown Lake). However, this does not mean that spawning was necessarily so widely spaced. Data collected in 1987 suggested that spawning may be separated by less than 2 weeks depending on the macroclimate and local conditions. Windless widespread warm conditions in 1987 caused atypical early spawning in several northern areas of Ontario, such as in Killarney Park (E. Snoucins, Ministry of Natural Resources, personal communication). Thus the conditions following spawning are most important for hatching of yellow perch eggs because local conditions and microhabitat of the eggs will affect their hatch time and their subsequent summer growth. This is in accord with growth and survival studies on young smallmouth bass, which showed increased survival with early hatch (Shuter et al. 1980).

For some freshwater species, first daily ring formation has been associated with swim-up stage. Graham and Orth (1987) reared smallmouth bass *Micropterus dolomieui* at 17–23°C and found that swim-up and first ring formation corresponded and occurred 5–8 d posthatch. However, considerable variation in ring number was also associated with the formation or deposition of microzones at this time of development. Taubert and Coble (1977) found that several species of sunfish (*Lepomis* spp.) formed their first daily ring on the day of swim-up. Miller and Storck (1982) concluded that the consistently countable daily ring formation in largemouth bass coincided with initiation of exogenous feeding. However, they observed 7–8 prolarval rings in largemouth bass *Micropterus salmoides* otoliths, formed daily between hatching and swim-up, that were visible for only 10–15 d after swim-up. On the other hand, Isely and Noble (1987) found that daily ring formation commenced at hatching in wild large-

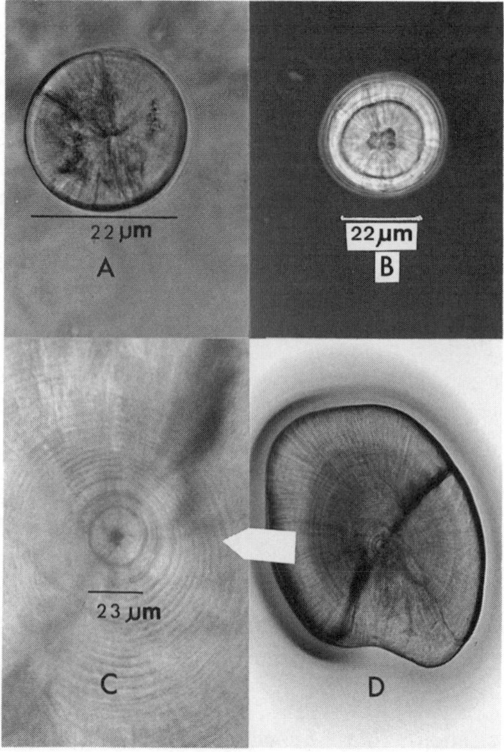

FIGURE 3.—**A.** Phase-contrast photomicrograph of yellow perch otolith with no daily rings (2 d before expected hatch). **B.** Otolith of yellow perch 2 d after hatch showing two daily rings, the second near the margin (dry, phase contrast). **C.** Central rings of an unground otolith from a 30-d-old yellow perch showing the distinct first daily microzone (dry, phase contrast). **D.** Same otolith as **C** showing whole structure (dry, phase contrast).

mouth bass. Their finding corresponds to our finding for yellow perch. Perhaps the prolarval rings Miller and Storck (1982) found in largemouth bass are not as clear as the early rings in yellow perch because the yolk-sac larvae of largemouth bass possess a larger yolk sac and the larvae are less active. We chose to count such rings, which were equally as clear as later rings in yellow perch. Our first posthatch ring was often very strong (Figure 3B) thus providing a benchmark. In fish over 3 cm total length, all central rings were easily observed if the otoliths were ground.

Incubation Period

A detailed account of ambient water temperatures is needed to estimate the actual spawning period (as opposed to hatch). Our preliminary studies indicate that fluctuating diel temperatures in situ tend to slow down the development below theoretical values obtained from constant temperature experiments. Our in situ eggs, however, were held in an atypically shallow warm bay, as opposed to the more typical incubation depth of 1–3 m, where ambient temperatures would be more stable. Under such conditions the laboratory rates of incubation should be more applicable.

There is considerable variation in the literature concerning incubation period or rate of development of yellow perch eggs. Guma'a's (1978) study for the closely related *Perca fluviatilis* indicated about 100 degree-days above 4.6°C (range 91–108 degree-days for 50% hatch). Hokanson and Kleiner (1974) reported an average of 120 degree-days for yellow perch, but did not estimate the minimum-temperature threshold. Utilizing Guama'a's minimum threshold of 4.6°C for our in situ incubation data, we obtained 107.4 degree-days (50% hatch) as opposed to Hokanson and Kleiner's 128–130 degree-days for constant laboratory conditions. Our working estimates of 1 week at 15°C or 6 d at 20°C (105–120 degree-days) was therefore probably within 1 d of the true incubation period.

Spawning Period

While the estimated hatch periods are useful, it would be desirable to predict the exact spawning date (start of incubation period) on the basis of light, temperature, and behavior. Knowing the length of the spawning season should provide better predictions on hatch mortality. Both Hokanson and Kleiner (1974) and Guma'a (1978) found higher mortalities at temperatures below 8°C and above 15°C with optimum survival between. Such factors could lead to better prediction of brood survival and could eventually be modeled.

Acknowledgments

We thank the following people for technical assistance in preparing this study: Mary Baker, Mary Boyd, Curtis Lewis, Steve Gardiner, Barb Mountney, Marilyn Metcalf, and Scott Finucan. Partial support came from Powles' National Sciences Engineering and Research Council grant A2353. Facilities were kindly provided to the senior author by the National Marine Fisheries Service, Beaufort Laboratory, whose staff were very helpful. We are grateful also to Tom Stewart who helped with collections; to James Nighswander who kindly provided boats and space at his cottage on Clear Lake; and to the referees who made many helpful suggestions.

References

Brothers, E. B., D. M. Williams, and P. F. Sale. 1983. Length of larval life in twelve families of fishes at "One Tree Lagoon", Great Barrier Reef, Australia. Marine Biology (Berlin) 76:319–324.

Campana, S. E. 1984. Microstructural growth patterns in the otoliths of larval and juvenile starry flounder, *Platichthys stellatus*. Canadian Journal of Zoology 62:1507–1512.

Fives, J. M., S. M. Warlen, and D. E. Hoss. 1986. Aging and growth of larval bay anchovy, *Anchoa mitchilli*, from the Newport River estuary, North Carolina. Estuaries 9:362–367.

Geffen, A. J. 1987. Methods of validating daily increment deposition in otoliths of larval fish. Pages 223–240 *in* R. C. Summerfelt and G. E. Hall, editors. Age and growth of fish. Iowa State University Press, Ames.

Graham, R. J., and D. J. Orth. 1986. Effects of temperature and streamflow on time and duration of spawning by smallmouth bass. Transactions of the American Fisheries Society 115:693–702.

Graham, R. J., and D. J. Orth. 1987. Otolith aging of young-of-year smallmouth bass. Pages 483–491 *in* R. C. Summerfelt and G. E. Hall, editors. Age and growth of fish. Iowa State University Press, Ames.

Gregory, R. S. 1983. Seasonal succession of larval fishes in a highly macrophytic lake, using a light-trap. Master's thesis. Trent University, Peterborough, Canada.

Gregory, R., and P. M. Powles. 1985. Chronology, distribution, and sizes of larval fish sampled by light traps in macrophytic Chemung Lake. Canadian Journal of Zoology 63:2569–2577.

Guma'a, S. A. 1978. The effects of temperature on the development and mortality of eggs of perch, *Perca fluviatilis*. Freshwater Biology 8:221–227.

Hokanson, K. E. F., and C. F. Kleiner. 1974. Effects of constant and rising temperature on survival and development rates of embryonic and larval yellow perch, *Perca flavescens* (Mitchill). Pages 437–448 *in* J. H. S. Blaxter, editor. The early life history of fish. Springer-Verlag, New York.

Houde, E. D. 1969. Sustained swimming ability of larval walleye (*Stizostedion vitreum vitreum*) and yellow perch (*Perca flavescens*). Journal of the Fisheries Research Board of Canada 26:1647–1659.

Isley, J. J., and R. L. Noble. 1987. Validation of daily ring deposition in otoliths of wild young-of-the-year largemouth bass. Texas Journal of Science 39:273–277.

Mansueti, A. J. 1964. Early development of the yellow perch, *Perca flavescens*. Chesapeake Science 5:46–66.

Miller, S. J., and T. Storck. 1982. Daily growth rings in otoliths of young-of-the-year largemouth bass. Transactions of the American Fisheries Society 111:527–530.

Miller, S. J., and T. Storck. 1984. Temporal spawning distribution of largemouth bass and young-of-year growth determined from daily otolith rings. Transactions of the American Fisheries Society 113:571–578.

Raney, E. C. 1959. Some young freshwater fishes of New York. New York State Conservationist 14(1): 22–28.

Ross, J., P. M. Powles, and M. Berrill. 1977. Thermal selection and related behaviour in larval yellow perch (*Perca flavescens*). Canadian Field-Naturalist 91:406–410.

Shuter, B. J., J. A. MacLean, F. E. J. Fry, and H. A. Regier. 1980. Stochastic simulation of temperature effects on first-year survival of smallmouth bass. Transactions of the American Fisheries Society 109:1–34.

Taubert, B. C., and D. W. Coble. 1977. Daily rings in otoliths of three species of *Lepomis* and *Tilapia mossambica*. Journal of the Fisheries Research Board of Canada 34:332–340.

Thorpe, J. 1977. Synopsis of biological data on perch, *Perca fluviatilis* Linnaeus, 1758, and *Perca flavescens* Mitchill, 1814. FAO (Food and Agriculture Organization of the United Nations) Fisheries Synopsis 113.

Warlen, S. M. 1982. Age and growth of larvae and spawning time of Atlantic croaker in North Carolina. Proceedings of the Annual Conference of Southeastern Association of Fish and Wildlife Agencies 34:204–214.

Zwanenburg, K. C. T. 1981. The relationship of growth and larval mortality rates to food availability as measured *in situ* for larval yellow perch (*Perca flavescens*). Master's thesis. McGill University, Montreal.

American Fisheries Society Symposium 5:68–79, 1988

Evidence for Baseline Flow Spikes as Spawning Cues for Colorado Squawfish in the Yampa River, Colorado

THOMAS P. NESLER

Fish Research Section, Colorado Division of Wildlife, Fort Collins, Colorado 80526, USA

ROBERT T. MUTH

Larval Fish Laboratory, Department of Fishery and Wildlife Biology
Colorado State University, Fort Collins, Colorado 80523, USA

ANTHONY F. WASOWICZ

U.S. Fish and Wildlife Service, Utah Cooperative Fisheries Research Unit
Utah State University, Logan, Utah 84321, USA

Abstract.—The Colorado squawfish *Ptychocheilus lucius* is an endangered species in the Colorado River basin. Understanding the spawning ecology of this cyprinid has been an important focus of research directed at recovery of the species. Comparison of yearly river flow regimes to estimated Colorado squawfish spawning periods in the Yampa River, Colorado, suggested that the first major flow spike in early summer, as river flows decline to near baseline, may act as an environmental cue that stimulates substantial spawning. In five cases during 1983-1986, estimated spawning peaks were associated with flow spikes of 27.5–71.4 m^3/s that occurred over a 2–3-d period sometime during late June to late July. These flow spikes represented 25–139% increases in river flow. Rainstorms representing 23–89% of total monthly precipitation appeared to cause the observed flow spikes. Spawning activity appeared to become negligible when flows decreased below 57–66 m^3/s. Flow spikes similar to those in 1983-1986 represented 34% of the 68 spikes documented for the Yampa River during the June–August period from 1934 to 1986. These flow spikes occurred in 19 of 53 years with a mean interval of 1.7 years. Seventy-four percent of the flow spikes of a magnitude similar to those in 1983–1986 peaked in 4d or less. Because of their recurrent nature, these flow spikes qualify as potential environmental cues serving to stimulate spawning by Colorado squawfish. These results may bear on recommended flow windows and water project operations criteria for the protection and management of this species in regulated river environments.

The Colorado squawfish *Ptychocheilus lucius* is a large piscivorous minnow endemic to the Colorado River basin. Presently, the Colorado squawfish is listed as endangered, and wild populations are found only in the upper basin in Colorado and Utah. Research on extant Colorado squawfish populations has been underway since the mid-1960s. Recent work has emphasized reproductive and early life habits, particularly as they relate to present-day physical and biological conditions in the system. One of the primary objectives of recovery plans for this endangered species is to identify and safeguard its habitat and life history requirements. In this paper, we examine the relationship between timing and extent of Colorado squawfish spawning (estimated from larval fish collections), and flow regimes and water temperatures in the Yampa River, Colorado, during 1980-1986, and discuss the role these physical variables may play as environmental cues that stimulate spawning activity by the species in this river. The Green–Yampa river system, which drains to the Colorado River, is considered to be one of the few remaining strongholds where spawning populations of wild Colorado squawfish still occur.

Successful reproduction by Colorado squawfish in the upper Colorado River basin was first verified by collection of young of the year from the Green River, Colorado–Utah, in 1964 (Vanicek and Kramer 1969). Evidence of successful reproduction by Colorado squawfish in the Yampa River was first verified by capture of larval fish in 1980 (Haynes et al. 1984). Subsequent radiotracking and sampling of adult Colorado squawfish have demonstrated that these fish migrate to and spawn in the lower 50 km of the Yampa River in Colorado (Figure 1) and in a 14.5-km reach of the Green River in Utah at Three Fords (Tyus et al. 1981; Wick et al. 1983; Tyus and McAda 1984; Tyus et al. 1987). The largest concentration of Colorado squawfish spawners in the Yampa River has been observed at river kilometers 24–30,

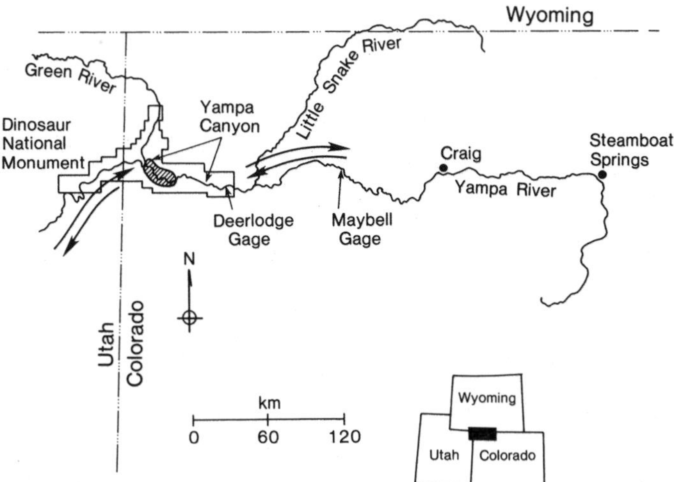

FIGURE 1.—Yampa River basin in Colorado. Hatched area denotes Colorado squawfish spawning area. Arrows denote migration routes of adult spawners from upper Yampa River, Colorado, and Green River, Utah.

measured from the river's mouth (Tyus et al. 1987). This river reach is considered the primary spawning ground in the Yampa River. Some Colorado squawfish have migrated 100–350 km round-trip via two or three rivers to use the spawning grounds on the Yampa and Green rivers (Wick et al. 1983; Tyus et al. 1987). Adjunct to these studies on adult spawners, intensive sampling for larval Colorado squawfish was conducted, especially downstream of suspected spawning areas and during the period of suspected spawning dates. In most cases, collections of Colorado squawfish larvae supported observations of presumed adult spawning behavior.

Documentation that Colorado squawfish larvae drift downstream was first achieved with the capture of two protolarvae by drift-net sampling in the Colorado River at Black Rocks in Ruby Canyon in 1982 (Haynes et al. 1985). The occurrence of Colorado squawfish as drifting larvae in the Yampa River (Yampa Canyon) was subsequently documented in 1983 (Haynes et al. 1985). The 1983 data indicated that the hatched larvae drifted out of the Yampa River and into the Green River. During 1983–1985, drift-net sampling was developed as a technique for sampling Colorado squawfish larvae (Haynes et al. 1985).

Capture data and total length of Colorado squawfish larvae collected by both drift net and seine have been used to estimate spawning dates for Colorado squawfish in the Yampa River (Haynes et al. 1984; Haynes et al. 1985; Nesler 1986). In this paper, these dates are related to Yampa River flow regimes and temperature data to identify the ranges of these environmental variables that occur during the estimated spawning period of wild Colorado squawfish in this relatively unregulated river.

Methods

From 1980 through 1984, habitats with low-velocity water flows (e.g., shallow shorelines, backwaters, and embayments) throughout the lower 33 km of the Yampa River were sampled with seines and dip nets for Colorado squawfish larvae (Table 1). Seine collections were made during post runoff periods during July and August; the seines had dimensions of 3.0 × 1.2 m and 1.0 × 1.2 m and 1.6-mm-square mesh. From 1983 through 1986, drift nets were used during July and August at a site on the Yampa River, 3.1 km above the Green River confluence, to collect larvae drifting from the primary spawning area 21 km further upstream. The nets were attached to fence posts, anchored in the substrate near shore at wading depth (1.4 m), and fished just below the water surface. These nets were 0.5-m-diameter, conical plankton nets mounted on 0.5 × 0.3-m rectangular steel frames, and fitted with 33-cm-long, removable polyvinyl chloride collection buckets with a threaded cod end (10-cm diameter). Each net had a 560-μm Nytex nylon mesh, a length of 4.0 m and an open-mesh-to-mouth area ratio of 11:1. Samples were taken at sunrise minus 0.5 h, noon, sunset plus 0.5 h, and midnight. Midnight samples were not taken in 1986. Sampling duration was 30 min to 2 h depending upon

TABLE 1.—Catches of Colorado squawfish larvae in the Yampa River, Colorado, 1980–1986.

Year	Number of samples	Capture dates	Capture locations (river km)	Number of larvae collected	Size range (mm[a])
		Seine and dip-net collections			
1980	215	Aug 23–25	0.2–13.9	46	14.0–29.0
1981	423	Jul 24–Aug 15	0.2–28.8	23	9.0–22.0
1982	295	Jul 7–25	0.5–19.6	20	9.9–21.0
1983	245	Jul 22–Aug 21	1.1–19.8	228	8.6–16.6
1984	148	Jul 26–Aug 6	0.2–25.4	249	8.1–15.1
		Drift-net collections			
1983	129	Jul 22–Aug 21	3.1	102	7.3–9.3
1984	168	Jul 17–Aug 8	3.1	89	7.2–9.5
1985	204	Jul 9–Aug 16	3.1	455	7.9–10.0
1986	198	Jul 15–Jul 31	0.1	124	8.4–11.0

[a]Total length.

the suspended debris load. Samples were returned to the Larval Fish Laboratory, Colorado State University, for processing. Colorado squawfish larvae were measured (total length) to the nearest 0.1 mm under a binocular dissecting microscope and assigned to a developmental phase (Snyder 1981). Further details of the sampling methodology were described by Haynes et al. (1985) and Nesler (1986).

Based on growth data provided by Hamman (1981), an age–growth relationship for hatchery-reared Colorado squawfish larvae during their first 107 d of life posthatching was developed by regression analysis (Haynes et al. 1985). These analyses showed that the age–growth relationship for Colorado squawfish larvae was best described in two distinct phases and resulted in two predictive equations for estimating posthatching age of individual larvae (Figure 2). The equation for larvae smaller than 22 mm total length was the principal one used because over 93% of the Colorado squawfish larvae collected were in this size range (Table 1). Hamman (1981) reported that the incubation time for Colorado squawfish eggs was 3.5–6 d at water temperatures of 20–22°C. For the purpose of predicting spawning time, 5 d were added to estimated posthatching ages to account for incubation time. The combined incubation time and larval posthatching age was used to back-calculate spawning time from the date of capture for individual larvae. Estimated spawning dates for individual larvae were aggregated in a frequency distribution that demonstrated beginning and ending dates and peak periods of spawning.

Comparisons of Colorado squawfish spawning periods with Yampa River flow regimes were made for each year from 1980 through 1986. Daily water discharge records from June 16 through August 15 for the Yampa River were compiled for each year from U.S. Geological Survey (USGS) data collected at the Maybell (1934–1986) and Deerlodge (1982–1986) gages (USGS 1934–1986;

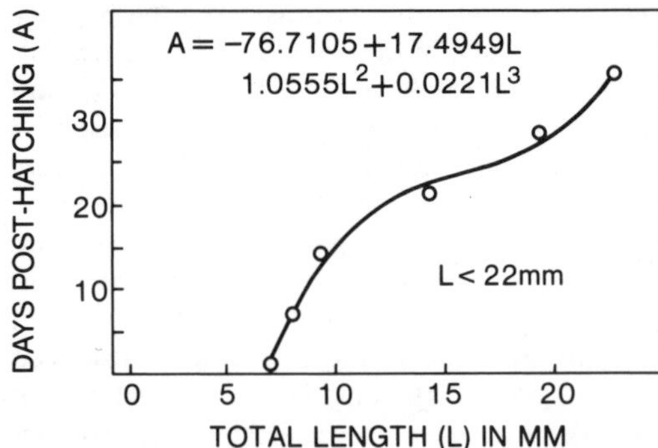

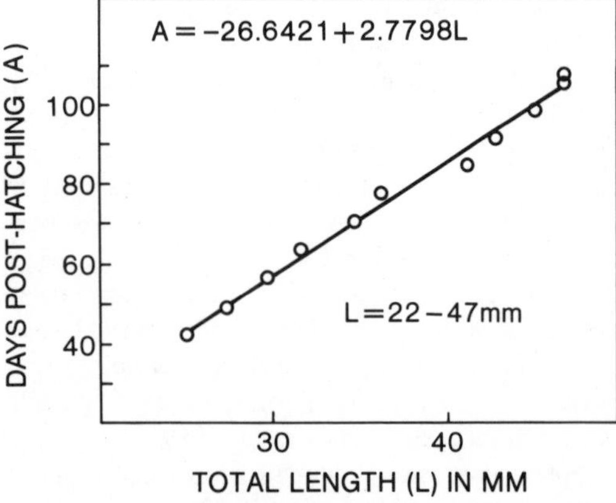

FIGURE 2.—Relationships of total length (L, ± 0.1 mm) to posthatching age for young-of-the-year Colorado squawfish. The upper curve applies to larval fish less than 22.0 mm long; the lower curve applies to fish of 22.0–47.0 mm.

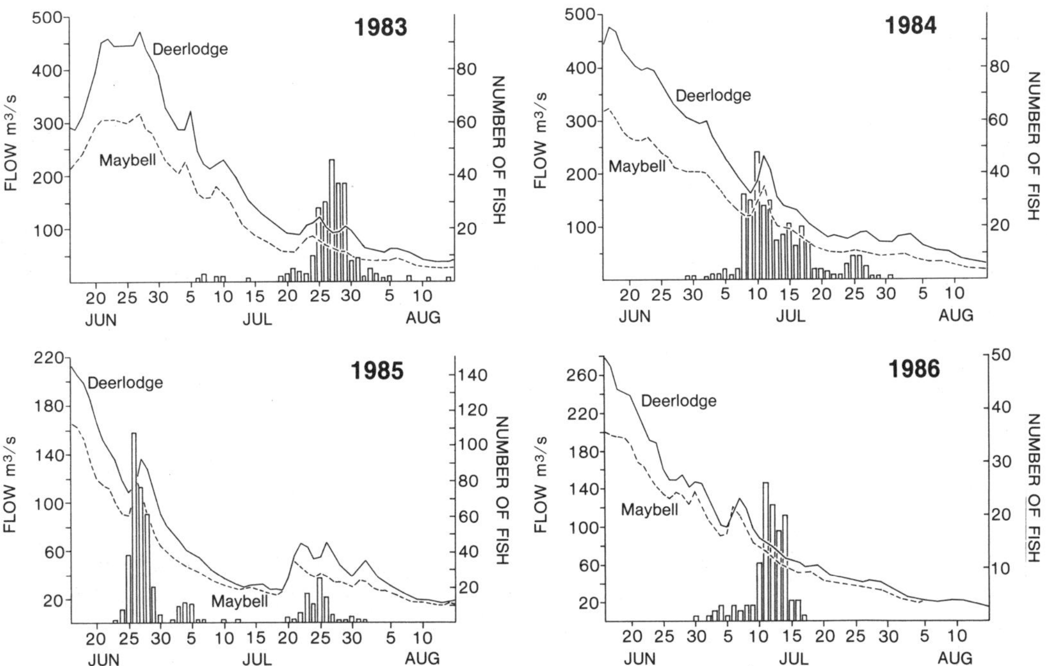

FIGURE 3.—Relationships of estimated Colorado squawfish spawning dates (vertical bars) to Yampa River flow regimes (curves) at the U.S. Geological Survey's Maybell and Deerlodge gages for 1983–1986. Number of fish represents number sampled and distributed according to estimated spawning date.

Figure 1). Months of record were selected to bracket the known Colorado squawfish spawning seasons as reported in existing literature. The years of record selected represented the available flow data for the Yampa River and were used to characterize the historical flow regime during the spawning season. Water temperatures were also obtained from USGS data collected at the Maybell gage. Precipitation data for the region including Dinosaur National Monument, Colorado, were compared to the Yampa River flow regime during June–August, 1980–1986, to determine the relationship between rainfall and flow spike events. Daily rainfall data were provided in NOAA (1980–1985, 1986).

Results

The frequency distribution of Colorado squawfish spawning dates for each year during 1983–1986 demonstrated prominent peaks in spawning activity (Figure 3). In 1983, peak spawning activity began July 24–25 and was closely related to the flow spike occurring July 22–25. Flows at Deerlodge increased from 89.7 to 122 m^3/s. Peak spawning activity in 1984 began July 8, and was associated with the flow spike occurring July 9–11. Flows at Deerlodge increased from 162.1 to 233.5 m^3/s. In 1985, two peaks in spawning activity were evident and appeared closely related to two major flow spikes. The first peak in spawning activity began June 25, and was associated with the flow spike occurring June 25–27. Flows at Deerlodge increased from 108.5 to 136 m^3/s. The second peak in spawning activity began July 23, and was associated with the flow spike occurring July 19–22. Flows at Deerlodge increased from 27.6 to 66 m^3/s. In 1986, peak spawning activity began July 10, following the flow spike occurring July 5–7. Flows at Deerlodge increased from 98.9 to 129.2 m^3/s. In these five cases, spawning peaks were associated with flow spikes of 27.5–71.4 m^3/s occurring over a 2–3-d period. These spikes represented 25–139% increases in river flow.

No prominent spawning peaks by Colorado squawfish were evident during 1980–1982 (Figure 4). In these 3 years, only 20–46 larvae were collected, compared to 124–455 in 1983–1986 (Table 1). As a result, only the range of spawning dates may be compared to the flow regime. In 1980 and 1981, onset of spawning appeared to be loosely associated with the occurrence of flow

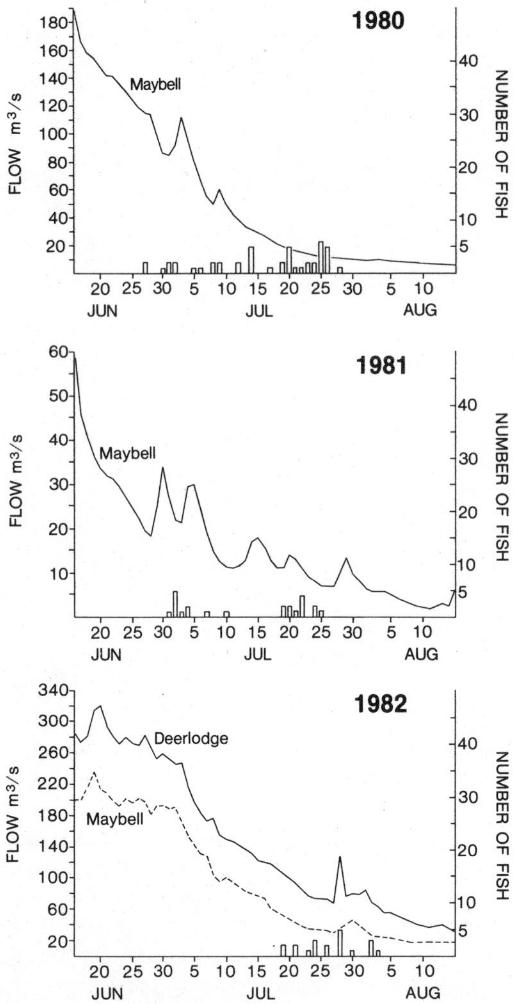

FIGURE 4.—Relationships of estimated Colorado squawfish spawning dates (vertical bars) to Yampa River flow regimes (curves) at the U.S. Geological Survey's Maybell and Deerlodge gages for 1980–1982. Number of fish represents number sampled and distributed according to estimated spawning date.

spike events, whereas in 1982, the flow spike occurred in the middle of the spawning period.

Comparison of the Deerlodge and Maybell hydrographs for the Yampa River each year from 1982 through 1986 (Figures 3 and 4) indicated that flow fluctuations were regional in their occurrence. The two USGS gages are 65 km apart, and it is presumed from this that similar flow fluctuations would also be evident at the Colorado squawfish spawning area 47 km downstream from Deerlodge. During 1982–1986, the flow spikes of interest began simultaneously at both sites but their peaks were simultaneous at both sites only in 1984. In 1983, 1985, and 1986, the flow spike peak occurred at the Deerlodge site 1 d later than at the Maybell site. Therefore, date of occurrence of a flow spike at the primary Colorado squawfish spawning area at km 24–30 was considered temporally proximal to its occurrence at the upstream gage stations. Flow spikes were associated with major rainstorms in the area of Dinosaur National Monument. In each year from 1980 to 1986, rainstorms representing 17–89% of the month's precipitation occurred immediately preceding the flow spikes related to Colorado squawfish spawning.

Examination of daily flow hydrographs for the Yampa River during the June 16–August 15 period from 1934 through 1986 suggested that baseline flow spikes might qualify as a recurring environmental variable for a long-lived, riverine fish species such as Colorado squawfish. Sixty-eight flow spikes were recorded during this season over the 53-year period (Table 2). Flow spikes similar to those observed for 1983–1986 (25.5 m^3/s or greater) represented 34% of the number documented and occurred in 19 of 53 years. The mean interval between these flow spikes was 1.7 years. The maximum interval was 8 years. The minimum magnitude of a flow spike needed to act as a spawning cue is unknown, so smaller flow spikes than those observed in 1983–1986 may be sufficient. Flow spikes of 14.2 m^3/s or greater represented 60% of the total number, and occurred in 32 of 53 years. The mean interval between occurrences was 0.6 years and the maximum interval was 5 years. The duration of these baseline flow spikes from beginning to peak was predominately 4 d or less (Table 3). Eighty-two percent of the 68 spikes peaked within this time frame. For flow spikes 25.5 m^3/s or greater, 74% peaked in 4 d or less. For flow spikes ranging from 14.2 to 22.7 m^3/s, 89% peaked in 4 d or less.

A relationship was also noted between flow level and cessation of notable spawning activity (Figure 3). Spawning activity in 1986 dropped markedly on July 14 when flow decreased to 66.3 m^3/s, and ceased when flow reached 58.4 m^3/s. Spawning activity during the first peak in 1985 appeared to drop to a negligible level by July 5, when river flow decreased to 57 m^3/s. During the second spawning peak in 1985, substantial spawning started on July 23, when flow reached 62.6 m^3/s after increasing from a low of 27.7 m^3/s. Spawning activity dropped off again after July 26 as flow decreased to 57.5 m^3/s. In 1984, peak spawning activity dropped off by July 19 when flow was at 95.8 m^3/s and spawning activity became negligible

TABLE 2.—Magnitude and relative frequency of baseline flow spikes in the Yampa River (Maybell gage) during June 16–August 15, 1934–1986.

Flow (m^3/s)	Number of occasions	Cumulative frequency (%)	Frequency (number/53 years)	Mean interval (years)	Maximum interval (years)
28.3+	18	26	17	2.1	8
25.5+	23	34	19	1.7	8
22.7+	25	37	19	1.7	8
19.8+	30	44	23	1.2	5
17.0+	36	53	28	0.8	5
14.2+	41	60	32	0.6	5
11.3+	50	74			
8.5+	59	87			
5.7+	67	99			
2.8+	68	100			

by July 28 when flow decreased to 79.6 m^3/s. In 1983, major spawning activity ceased by August 1 when river flow dropped below 62.9 m^3/s.

Discussion

Our hypothesis formed from these comparisons of flow spikes and spawning periods is that a flow spike near baseline in the Yampa River acts as a recurring environmental cue to adult Colorado squawfish to initiate the spawning process of egg deposition and fertilization. The flow spike represents the observable physical event we have associated with Colorado squawfish spawning, but we do not exclude the possibility that, in reality, one or more environmental variables coincidental to the flow spike may be serving as the actual cue perceived by the adult fish. Prior to a discussion of the role the flow spike cue might play in the reproductive ecology of this species, an evaluation of the approach used to establish the above hypothesis is necessary. The limited population size associated with an endangered species, the difficulties of representative sampling in the aquatic environment, and the indirect methods used complicate interpretations of the biological significance of Yampa River flow spikes in the spawning ecology of the Colorado squawfish. Although the cumulative affect of these sources of variation upon the alignment of the flow spike with peak spawning by Colorado squawfish is unpredictable, these uncertainties are considered to be responsible for the noise evident in the alignments and do not detract from an otherwise remarkably consistent relationship. The strength of the flow-cue hypothesis is based on the close association noted between flow spikes and peak spawning activity in 4 consecutive years.

Potential Sources of Error

Use of different sampling gear may be responsible for differences in the total catch of larvae between 1983–1986 and 1980–1982. The inclusion of drift-net sampling in 1983–1986 and the potential inadequacy of seine sampling for representatively sampling drifting Colorado squawfish larvae may account for the observed differences rather than a biological change signifying increased spawning success. Drift-net data from Haynes et al. (1985) and Nesler (1986) demonstrated that larval Colorado squawfish were transported downstream and out of the Yampa River within 3–15 d after hatching. The majority of Colorado squawfish larvae produced in the Yampa River in a given year may be transported downriver as a discrete group that is present in any 6-km segment of the river for only 1–2 d (Nesler 1986). Drift-net sampling over time at a permanent station would ultimately pick up this concentration of larvae and

TABLE 3.—Frequency distribution of flow spikes according to duration in number of days to peak.

Flow (m^3/s)	Number of spikes of duration						
	1 d	2 d	3 d	4 d	5 d	6 d	7 d
25.5+	2	5	5	5	1	3	2
14.2–22.7+	4	6	5	1	1	0	1
Total	6	11	10	6	2	3	3
All flows	8	21	16	11	4	5	3

reflect the temporal variation in number of drifting larvae. In contrast, seine sampling throughout a 75-km river reach over the 5–6 d required to float the reach by raft could easily miss this discrete group. In 1980–1982, seine sampling may have missed the concentration of larvae in the river. In 1983–1984, the seining catch per effort for larval Colorado squawfish was greater than in 1980–1982 (Haynes et al. 1985), and some portion of the peak production of larvae was sampled. However, it remains uncertain whether one-time seine sampling throughout a river reach can adequately quantify the transient pulse of drifting Colorado squawfish larvae in the Yampa River. It is presumed that sampling of drifting larvae and estimation of spawning dates were more representative in 1983–1986, when a combination of sampling techniques or drift-netting alone were used, than in 1980–1982, when seine sampling alone was used. Thus, the low catches of Colorado squawfish larvae in 1980–1982 might not indicate poor reproductive success.

In using the predictive age–growth equations, we assumed that growth and incubation time for wild Colorado squawfish stocks in the Yampa River were similar to those of hatchery stocks raised in Willow Beach National Fish Hatchery, Arizona. It is probable that there is greater year-to-year variability in both these variables among wild than among hatchery progeny. Another assumption represented by the polynomial age–growth equation (Figure 2) is that all larvae are approximately 6.7 mm total length at hatching. Size at hatching within a year class of wild progeny (as well as within hatchery stocks) is certainly variable. Hamman (1981) reported that hatching sizes of Colorado squawfish larvae varied from 6.0 to 7.5 mm total length; differences between the smallest and largest larvae were 0.5 to 1.0 mm for a given batch of eggs. Size differences of this range result in 3- to 5.5-d differences in posthatching age estimated with the polynomial equation. As pointed out earlier, the incubation times reported by Hamman (1981) show a 2.5-d range, or a 1–1.5-d variation about the mean incubation time of 5 d used here. These factors may apply to a few, some, or all of the larvae in a particular year.

On the positive side, mean water temperatures at the Yampa Canyon spawning ground ranged from 22 to 23.5°C in 1981–1984 (Tyus et al. 1987). This compares favorably to the 20–24°C water temperatures present at the hatchery during the Hamman (1981) study upon which the larval age–growth equations were based. Also, Hamman (1981) reported that cold river water was added periodically to the recirculated-water raceways holding the Colorado squawfish larvae, lowering the water temperature from 23–24°C to 17–18°C. Whether intentional or not, this action was probably a close simulation of water temperature fluctuations that occur in the Yampa River. Because growth data used in the predictive equations were averaged over all rearing conditions reported in Hamman (1981), the age–growth equations probably provide a fair approximation of growth conditions for wild Colorado squawfish larvae.

Colorado squawfish collected by drift net in Yampa Canyon were predominately protolarvae (7.3–9.0 mm total length) whose nutritional needs were still supplied from a yolk sac. Thus, varying food resources in the wild for drifting Colorado squawfish larvae would be of limited concern with regard to the validity of the age–growth equations. However, the effect of a variable environment upon the growth of actively feeding mesolarval Colorado squawfish presents another unknown variable. For example, the comparison of the 1984 spawning peak with the flow regime suggested that the flow spike occurred 1 d later than the start of peak spawning activity. This incongruity in alignment may be a function of the predominant influence of seine-collected larvae, which provided the basis of the 1984 spawning peak on July 8–12. Colorado squawfish larvae collected by seines were typically larger than those captured by drift nets in 1983 and 1984 (Table 1). It is possible that some of the seine-collected larvae were actively selecting habitats with low-velocity water that was warmer than in the main channel, and grew faster than would be predicted. Earlier spawning dates would be calculated from these larger larvae than actually occurred. Valdez et al. (1985) showed that, for native cyprinids in the Colorado River, smaller larvae usually appeared in drift nets and larger larvae were found along shorelines sampled by seines.

Extrapolation of flow patterns from USGS gages distant from the Colorado squawfish spawning area suggested another source of variation. Comparison of the Maybell and Deerlodge hydrographs demonstrated very similar patterns, but differences in the timing of peak spike flows between the two gage sites suggested that a spike peak arrives at primary spawning area in Yampa Canyon 1 d after it passes the Deerlodge gage. Of greater importance to the assumption of similar flow patterns between the gages and the spawning area is that the beginning of the flow spike occurred simultaneously at the two gage sites, suggesting that rainstorms affected the region and not a more limited upstream locale.

Sampling Artifact or Spawning Cue?

The frequency distribution of estimated spawning dates was a function of number of larvae sampled. The question arises as to whether peaks observed in these frequency distributions in 1983–1986 were sampling artifacts caused by a flushing effect in the spawning grounds due to the flow spike or a result of drift nets sampling more water (and thus more larvae) during rising flow even though catch per unit of water volume sampled remained the same. The answers are no to both, because lengths, not numbers, of captured larvae were used to estimate spawning dates, that is, when the eggs were deposited in the gravel. The bimodal distribution of Colorado squawfish larvae captured in 1985 (Nesler 1986) supports this argument. The length range of Colorado squawfish larvae responsible for the July 23 spawning peak was similar to that of larvae responsible for the June 25 spawning peak (8.0–9.5 mm versus 7.9–9.4 mm total length, respectively). This indicates that both groups consisted of protolarvae and that the later spawning peak was not an artifact caused by older, larger larvae being flushed out with the second flow spike.

Hamman (1981) observed that Colorado squawfish eggs are adhesive and that hatched larvae do not leave the interstitial spaces of the cobble substrate for 3–4 d following hatching. These observations suggest that the eggs are fairly well protected from flow variations and that Colorado squawfish larvae would emerge from the gravel some time after the flow spike occurrence. Larvae were captured at lower flow levels well after the flow spike event occurred. Also, larvae were captured by drift nets in the nearshore zone of the river cross section. The characteristic low flow of this zone is unaffected by the amount of water moving downstream. The nearshore zone, and hence the drift-net sampling site, merely changed its relative position in the channel with little increase in current velocity or volume of water sampled. The critical point is that estimated spawning dates derived from total lengths of the captured larvae indicate the larvae were deposited in the gravel as eggs during a flow spike event.

The spawning periods estimated from collections of larval Colorado squawfish have been corroborated by radiotracking data on adults and collections of ripe adults on the spawning ground. Estimates of spawning dates made from larval collections, radiotracking, and collection of ripe adults were generally in agreement during 1981–1985 (Tyus et al. 1987). Estimations of optimal spawning periods for 1981–1985 were presented by Tyus et al. (1987). Their estimates and ours agree closely except for 1984; their estimated optimal spawning period was July 20–August 10 in 1984, but we estimated the end of peak spawning activity at July 18. The optimal spawning period estimated by Tyus et al. (1987) for 1984 was based on data from five radiotracked fish, the collection of 16 ripe adults (only two of which were females), and 58 larvae collected in the Green River below the confluence of the Yampa River in late August–early September. Our estimate resulted from 365 larval Colorado squawfish collected from July 17 through August 8. Tyus et al. did observe ripe Colorado squawfish adults on the primary spawning reach in the Yampa River at least as early as July 15. We believe that substantial spawning activity occurred prior to this date but was not detected from radiotracking observations or adult fish sampling efforts by Tyus et al. (1987).

In reference to flow conditions at the cessation of peak spawning activity, remarkably similar flows occurred in 1983, 1985, and 1986, whereas the 1984 flow seemed notably higher. Another unaccountable variable was the overall number of adult Colorado squawfish participating in the spawning run from year to year. Most of the available adult Colorado squawfish might have participated during the peak of spawning in 1984, leaving relatively few adults to spawn afterwards even though flows were still above 55–65 m^3/s. The cessation and resumption of spawning with fluctuating flows in 1985 supported the hypothesis that major spawning activity ceased at flows below this threshold.

Flow Spikes as Spawning Cues

Our approach to the observed relationship between Colorado squawfish spawning and seasonal flow regimes in the Yampa River is quite simplistic in scope relative to the complex interaction of external and internal cues that may initiate and regulate gonadal development, migratory behavior, spawning, and egg deposition. The relationship between internal and external rhythms is subtle and the timing of a species' reproductive cycle is a compromise involving many environmental considerations (Bye 1984). Environmental factors serve as cues synchronizing the internal reproductive cycles of the individuals in a population (Lam 1983). Timing of annual spawning has evolved to ensure that the optimal environmental conditions are present and that young hatch and commence feeding at the optimum season for survival (Nikolsky 1963; Schwassman 1971; Lam 1983; Bye 1984). A temporally appropriate response to external cues maximizes reproductive success by achieving an optimal balance between survival of adult spawners and progeny (Stacey 1984). Bye (1984) suggested that any recurring environmental variable within the sensory competence of an organism can act as a timing cue. To extend this concept further, multiple environmental cues may be integrated in a fish species' reproductive cycle, either interactively or in sequence, to synchronize gonad maturation (gametogenesis), migration, and spawning (Liley 1969; Lam 1983; McKeown 1984). A general consensus is evident in the literature that photoperiod and temperature interact as important regulatory factors in the timing and process of gonad maturation in many fish species and especially cyprinids (DeVlaming 1972; Magnuson et al. 1979; Lam 1983; Bye 1984; McKeown 1984; Stacey 1984). However, both Lam (1983) and Stacey (1984) indicated that although these environmental factors ensure that the gonads are mature at the appropriate season, other specific stimuli may be required to initiate the final phase of gonad maturation (ovulation and spermiation) and the release of gametes. Potential stimuli include current velocity, water quality, substrate, barometric pressure, or pheromone release due to aggregation of potential mates.

Vanicek and Kramer (1969), Wick et al. (1983), Haynes et al. (1984, 1985), Tyus and McAda (1984) and Tyus et al. (1987) have concluded that Colorado squawfish spawn as river flows decrease in early summer and that the timing may be largely influenced by water temperatures. Threshold temperatures of 20–22°C and the number of degree-days above 18°C have been proposed as key factors. This conclusion is supported by evidence from hatchery operations. Both Toney (1974) and Hamman (1981) reported temperatures of 20–22°C as important for spawning. Observations by Hamman (1981) under hatchery conditions indicated that Colorado squawfish spawned only after injection with gonadotropin when the temperature was 18°C, but they spawned spontaneously at temperatures of 20°C and above.

Evidence that temperature is important for Colorado squawfish spawning is substantial, but we propose a different perspective of its role. The water temperature data in Haynes et al. (1985) indicate that some spawning activity in 1983–1984 occurred when temperatures in the Yampa River were less than 20°C. In 1985, the mean temperature during peak spawning was only 16.1°C (Tyus et al. 1987). Temperature data at the Maybell gage showed that a drop of 4°C (18 to 14°C) occurred on June 25–27, concurrent with the flow spike and peak spawning activity by the Colorado squawfish (Nesler 1986). This association of events was clearly divergent from the spawning temperature threshold hypothesis that had been formulated prior to 1985. This relationship was also directly opposite that observed by Beamesderfer and Congleton (1981) for northern squawfish *Ptychocheilus oregonensis* in the St. Joe River, Idaho. They reported a dramatic decline in adult numbers and spawning activity coinciding with a 4°C drop in river temperature (15 to 11°C) due to a rainstorm that raised the river level approximately 22%. The estimated period of peak spawning in the Yampa River in 1985 occurred at water temperatures under which hatchery fish would require injection to induce spawning, which also supports the flow-cue hypothesis. Our data suggest that temperature may fulfill a secondary or supplementary role in the spawning requirements of Colorado squawfish in the Yampa River. Differences in spawning behavior between Colorado and northern squawfish may be related to stark differences in their respective habitats. Colorado squawfish have adapted to desert river ecosystems in more southerly latitudes receiving 25 cm annual rainfall, whereas northern squawfish have adapted to coldwater rivers in northern latitudes that receive over 77 cm annual precipitation.

Some spawning was calculated to dates that preceded any flow cue. Several explanations may account for these early outliers. Relative to assumptions made earlier, wild progeny may experience variable environmental conditions that affect growth, and their growth curve may be unlike the hatchery-based growth curves used here. From a genetic perspective, it is likely that a segment of each year class of Colorado squawfish larvae has inherent growth characteristics unlike those of Willow Beach hatchery fish. Further, not all adult fish may respond equally to an environmental cue for migration or spawning.

In discussions of environmental cues leading to ovulation and spawning, Stacey (1984) highlighted differences between salmonids and cyprinids.

This is particularly relevant because the spawning ecology of Colorado squawfish presents a cyprinid reproductive strategy that incorporates salmonid attributes. In salmonids, ovulation occurs after ovarian development is completed, and is analogous to spontaneous ovulation. Spawning migrations of salmonids have evolved as a reproductive tactic to synchronize gonadal maturation, ovulation, and egg deposition with suitable temperatures and flows, proper substrate quality, absence of predators, and so forth. In contrast, ovulation and spawning by cyprinids occurs rapidly in response to specific external factors relevant to reproductive success (Stacey 1984). Stacey et al. (1979) demonstrated that the presence of aquatic vegetation was an effective stimulus for ovulation and spawning by goldfish *Carassius auratus*, even at suboptimal water temperature. In the absence of vegetation, goldfish spawned only when the preferred temperature threshold was reached. Colorado squawfish reproduction displays a mixture of these attributes, placing the species somewhere in the middle of the ecological continuum between cyprinids and salmonids.

How does the flow cue hypothesis fit with the reproductive strategy of the Colorado squawfish? Based on the accumulated works and ideas of numerous researchers in the Colorado River basin, the following scenario may be developed. In the spring, adult Colorado squawfish respond to one or more specific cues, such as increasing flow and warm water temperatures in backwater habitats, and begin gonad maturation and migratory movements. From radiotracking studies, Wick et al. (1983) hypothesized that Colorado squawfish using backwater habitat in the spring were stimulated by the warmer water, which accelerated gonad maturation and triggered migration. The spring runoff flow and temperature pattern may accelerate maturation and migration processes. As adult fish continue their spawning migration, other cues such as olfactory stimuli from water quality changes and groundwater seepage may aid the fish in locating the appropriate migration route and spawning ground. Observations of the movements of downstream-migrating Colorado squawfish around the primary spawning ground in Yampa Canyon by Wick et al. (1983) and characterization of this spawning area as one of sandstone–limestone seeps by Tyus (1985), led to an adaptation of Harden-Jones's (1981) groundwater seepage hypothesis, whereby migrating Colorado squawfish may locate the spawning ground by olfactory detection of water sources on which they had imprinted. Arriving at the spawning ground, the fish first aggregate in quiet pools, then spawn in shallower, faster current over cobble-gravel substrate (Tyus et al. 1987). The observed transition between these behaviors and habitat types is abrupt. In the lower Yampa River, a spike in the baseline flow is the cue for ovulation and spawning by the aggregated spawners. In the event a flow spike of appropriate magnitude does not occur, an ultimate increase in water temperature past the 20–22°C threshold may stimulate spontaneous ovulation and spawning.

Historically, rivers of the Colorado River basin have had great seasonal fluctuations in discharge and associated variations in temperature, turbidity, dissolved solids, sediment transport, and allocthonous material (Wick et al. 1983; Tyus et al. 1987; Carlson and Muth in press). The reproductive migrations and larval drift displayed by Colorado squawfish in the Green–Yampa subbasin have evolved in this rigorous environment (Wick et al. 1983). These long-range, potamodromous migrations by both adult and larval fish are matched by few other North American cyprinid species (Tyus 1986; Tyus et al. 1987), but the adaptive importance of migration and larval drift in the reproductive ecology of fishes has been well established. Spawning migrations optimize reproductive success through the aggregation of suitable mates under favorable spawning conditions and the placement of newly hatched young in an environment relatively free of predators (Nikolsky 1963; Northcote 1967; McKeown 1984). Downstream displacement of young fish, either through passive drift or active migration, is considered adaptive because it carries the progeny to productive feeding areas, enhancing growth and thereby further limiting the vulnerability of the young fish to predation (Nikolsky 1963; McKeown 1984). Stream drift takes larval Colorado squawfish spawned in the lower Yampa River into the more abundant backwaters of the Green River in late summer, when warmer water temperatures enhance food production and fish growth in these nursery habitats (Tyus and McAda 1984; Haynes et al. 1985; Tyus 1986). Tyus (1986) suggested that spawning by Colorado squawfish is timed to occur at the end of the flood season to exploit river transport as a means to disperse progeny and to coincide with the formation of the productive nursery backwaters downstream from spawning areas. It is plausible to suggest that the flow spike cue, in a background of decreasing flow and increasing water temperature, signals the approach of baseline flow conditions and the emergence of favorable backwater habitat downstream.

Colorado Squawfish Spawning and Water Development

Flow and temperature are known key elements in the spawning requirements of Colorado squawfish. Our hypothesis is based on indirect evidence linking estimated spawning activity and flows, but verification of this and other requirements of the spawning populations of Colorado squawfish may not be possible until the species' status improves. Considerable effort has been expended to establish flow requirements for various life history stages of the Colorado squawfish. Depth and velocity curves describing habitat suitability are being constructed for five life stages of the Colorado squawfish including spawning. These recommendations will guide continuing recovery programs and consultation requirements for water development projects. An important consideration that must be acknowledged in these recommendations is that the native fish fauna has adapted to an environment of cyclical patterns. However variable the absolutes, such as maximum spring flow, the pattern of change and seasonal flux tends to occur regularly and predictably. It is logical to assume that species adapted to such an environment also become more or less dependent upon these cycles and fluxes. Hubbs (1972) suggested that a steady-state environment is deleterious to most species of native stream faunas and floras. Few streams can be characterized as stenotypic environments, so any manipulation of a stream that dampens environmental flux may be considered less favorable for the native biota. Colorado squawfish have been observed to migrate and spawn in a considerable range of flows since 1980. Holden (1980) and Haynes et al. (1985) concluded, from evidence of relative yearly spawning success and year-class abundance that high spring flows were necessary to create and maintain suitable spawning habitat. Tyus (1986) hypothesized that the Colorado squawfish, as a large, long-lived riverine species, relies on the production of many progeny during wet years.

Our results suggest that flow fluctuations during late June to early August may have important influences upon successful spawning by Colorado squawfish. These results may also have a direct bearing upon recommended operations criteria for water projects in order to protect this species in regulated river environments. It may be necessary to simulate natural flow patterns containing spring runoff peaks and baseline flow spikes to stimulate successful spawning by the Colorado squawfish in downstream spawning areas.

Acknowledgments

The data base used here was collected during several studies conducted by the Colorado Division of Wildlife. These studies have been funded through Colorado's Nongame Wildlife Income Tax checkoff program and through Section-6 funding from the U.S. Fish and Wildlife Service under the authority of the Endangered Species Act of 1973. We acknowledge the helpful ideas contributed by colleagues within the interagency coalition of the Colorado River Fishes Project in the Upper Basin, members of the Desert Fishes Council, and participants in the 11th Larval Fish Conference.

References

Beamesderfer, R. C., and J. L. Congleton. 1981. Spawning behavior, habitat selection, and early life history of northern squawfish with inferences to Colorado squawfish. University of Idaho, Idaho Cooperative Fishery Research Unit, Report 2, Moscow.

Bye, V. J. 1984. The role of environmental factors in the timing of reproductive cycles. Pages 187–205 *in* G. W. Potts and R. J. Wootton, editors. Fish reproduction: strategies and tactics. Academic, London.

Carlson, C. A., and R. T. Muth. In press. The Colorado River: lifeline of the American Southwest. Canadian Journal of Fisheries and Aquatic Sciences.

DeVlaming, V. L. 1972. Environmental control of teleost reproductive cycles: a brief review. Journal of Fish Biology 4:131–140.

Hamman, R. L. 1981. Spawning and culture of Colorado squawfish in raceways. Progressive Fish-Culturist 43:173–177.

Harden-Jones, R. F. 1981. Fish migration: strategy and tactics. Society for Experimental Biology Seminar Series 13:139–165.

Haynes, C. M., T. A. Lytle, E. J. Wick, and R. T. Muth. 1984. Larval Colorado squawfish (*Ptychocheilus lucius* Girard) in the upper Colorado River basin, Colorado, 1979–1981. Southwestern Naturalist 29:21–34.

Haynes, C. M., R. T. Muth, and T. P. Nesler. 1985. Identification of habitat requirements and limiting factors for Colorado squawfish and humpback chub. Colorado Division of Wildlife, Job Final Report SE-3-4, Fort Collins.

Holden, P. B. 1980. The relationship between flows in the Yampa River and success of rare fish populations in the Green River system. BioWest, Inc., PR-31-1, Logan, Utah.

Hubbs, C. 1972. Some thermal consequences of environmental manipulations of water. Biological Conservation 4:185–188.

Lam, T. J. 1983. Environmental influences on gonadal activity in fish. Pages 65–116 *in* W. S. Hoar, D. J. Randall, and E. M. Donaldson, editors. Fish physiology, volume 9, part B. Academic, New York.

Liley, N. R. 1969. Hormones and reproductive behavior in fishes. Pages 73–116 *in* W. S. Hoar and D. J. Randall, editors. Fish physiology, volume 3. Academic, New York.

Magnuson, J. J., L. B. Crowder, and P. A. Medvick. 1979. Temperature as an ecological resource. American Zoologist 19:331–343.

McKeown, B. A. 1984. Fish migration. Timberline, Portland, Oregon.

Nesler, T. P. 1986. Aquatic nongame research—1985–86. Squawfish–humpback studies. Colorado Division of Wildlife, Annual Job Progress Report SE-3, Fort Collins.

Nikolsky, G. V. 1963. The ecology of fishes. Academic, London.

NOAA (National Oceanic and Atmospheric Administration). 1980–1985. Hourly precipitation data, volumes 30–35, numbers 6–9, Washington, D.C.

NOAA (National Oceanic and Atmospheric Administration). 1986. Hourly precipitation data, volume 36, Washington, D.C.

Northcote, T. G. 1967. The relation of movements and migrations to production in freshwater fishes. Pages 315–345 *in* S. D. Gerking, editor. The biological basis of freshwater fish production. Blackwell Scientific, Oxford, England.

Schwassman, H. O. 1971. Biological rhythms. Pages 371–428 *in* W. S. Hoar and D. J. Randall, editors. Fish physiology, volume 6. Academic, New York.

Snyder, D. E. 1981. Contributions to a guide to the cypriniform fish larvae of the upper Colorado River system in Colorado. United States Bureau of Land Management, Biological Sciences Series 3, Denver.

Stacey, N. E. 1984. Control of the timing of ovulation by exogenous and endogenous factors. Pages 207–222 *in* G. W. Potts and R. J. Wootton, editors. Fish reproduction: strategies and tactics. Academic, London.

Stacey, N. E., A. F. Cook, and R. E. Peter. 1979. Spontaneous and gonadotropin-induced ovulation in the goldfish, *Carassius auratus* L.: effects of external factors. Journal of Fish Biology 15:349–361.

Toney, D. P. 1974. Observations on the propagation and rearing of two endangered fish species in a hatchery environment. Proceedings of the Annual Conference Western Association State Game and Fish Commissioners 54:252–259.

Tyus, H. M. 1985. Homing behavior noted for Colorado squawfish. Copeia 1985:213–215.

Tyus, H. M. 1986. Life strategies in the evolution of the Colorado squawfish (*Ptychocheilus lucius*). Great Basin Naturalist 46:656–661.

Tyus, H. M., R. L. Jones, and L. M. Trinca. 1987. Green River rare and endangered fish studies, 1982–1985. U.S. Fish and Wildlife Service, Colorado River Fishes Project, Final Report, Vernal, Utah.

Tyus, H. M., and C. W. McAda. 1984. Migration, movements and habitat preferences of Colorado squawfish, *Ptychocheilus lucius*, in the Green, White and Yampa rivers, Colorado and Utah. Southwestern Naturalist 29:289–299.

Tyus, H. M., E. J. Wick, and D. L. Skates. 1981. A spawning migration of Colorado squawfish in the Yampa and Green rivers, Colorado and Utah, 1981. Proceedings of the Desert Fishes Council 13:102–108, Bishop, California.

USGS (United States Geological Survey). 1934–1986. Water resources data, Colorado, water years 1934–1986, volume 3. USGS Water Data Reports CO-34-3 to CO-86-3, Lakewood, Colorado.

Valdez, R. A., J. G. Carter, and R. J. Ryel. 1985. Drift of larval fishes in the upper Colorado River. Proceedings of the Annual Conference Western Association of Fish and Wildlife Agencies and the Western Division American Fisheries Society 65:171–185.

Vanicek, C. D., and R. H. Kramer. 1969. Life history of the Colorado squawfish, *Ptychocheilus lucius*, and the Colorado chub, *Gila robusta*, in the Green River in Dinosaur National Monument, 1964–1966. Transactions of the American Fisheries Society 98: 193–208.

Wick, E. J., D. L. Stoneburner, and J. A. Hawkins. 1983. Observations on the ecology of the Colorado squawfish (*Ptychocheilus lucius*) in the Yampa River. Colorado Division of Wildlife, Endangered Wildlife Investigations SE-3-5; National Park Service Water Resources Field Support Laboratory Report 83-7, Fort Collins.

American Fisheries Society Symposium 5:80–88, 1988

Variation in Pigment and Nape Morphology of Larval Tidepool Sculpin

JEFFREY B. MARLIAVE

Vancouver Public Aquarium, Post Office Box 3232
Vancouver, British Columbia V6B 3X8, Canada

Abstract.—Key characters for the identification of yolk-sac larvae of the tidepool sculpin *Oligocottus maculosus* include the presence of a nape bubble and the count of postanal ventral midline (PVM) melanophores. However, both of these features vary among larvae of this species, limiting their taxonomic utility. In addition, previously undescribed pigment patterns on the ventral and lateral gut are polymorphic, being present or absent in different geographic regions and in different clutches. Precocial development of cranial pigment occurs by the time of hatching in some larvae. Clinal geographic variation occurs in counts of PVM melanophores. Certain combinations of these pigment and nape variants make identification of such larvae on the basis of previous descriptions very difficult. Unusual pigment patterns of reared larvae appear more likely attributable to limited sources of animals for culture than to artifacts of the culture environment.

Identification of fish larvae requires integration of diverse information, yet it ultimately involves the matching of available, recognizable similarities between specimens (Powles and Markle 1984). Such unique character states tend to be fewer among larvae than among adult fishes. Of the characters most often used for larvae, pigment is the only type that has been used for all early life history stages and at all taxonomic levels in systematic investigations (Kendall et al. 1984). Kendall et al. (1984) discussed the problems of using pigment as a taxonomic character, and they suggested that the most serious difficulties involve convergence of pigment patterns in unrelated species and the physiologically variable state of melanophore contraction. Although Powles and Markle (1984) indicated that regional variation in appearance of larvae can cause problems with identification, there appears to be little published documentation of polymorphic or clinal variation in pigment patterns or in other features indicated as diagnostic characters. One exception is a study of myomere-count variation among Mississippi River fish larvae (Bosley and Conner 1984). The pragmatic requirement for usable key characters to identify fish larvae appears to have resulted in rather broad acceptance of pigment patterns as conservative diagnostic characters, especially with marine species. Although generally correct, such acceptance causes problems when pigment variants exist within a species. Similar confusion can result from polymorphism of other types of diagnostic characters as well, especially when several of such features can vary.

The present study provides evidence for extreme variability in a number of the diagnostic characters cited for identification of yolk-sac larvae of the tidepool sculpin *Oligocottus maculosus*. Most of these variable characters are pigment patterns such as cranial, lateral, and ventral gut melanophores that had not been observed on specimens from the open Pacific coast from California to Washington (Stein 1973; Washington 1981). In addition, the unusual nape bubble described by Washington (1981) for preflexion tidepool sculpin larvae varies from being prominent on some larvae to absent on others. An analysis of these features indicates parental effects within a locality, as well as geographic patterns of variation.

Methods

Yolk-sac larvae were obtained from locations in southwestern British Columbia from the outer coast of Vancouver Island around to the inland waters surrounding the city of Vancouver (Figure 1). Eyed-egg masses were collected from the *Mytilus–Balanus* intertidal zone in most areas and incubated in running seawater. Larvae were killed in 1% formalin in seawater upon hatching, then fixed in 3% formalin. In certain localities, larvae were collected in plankton tows in rocky nearshore areas (Marliave 1986). Tows were limited to 3–5 min duration, so that larvae were alive on completion of a tow, then larvae were killed and fixed as for cultured ones. Field-caught larvae were sorted and identified with dissecting stereomicroscopes, which were also used for counts and measurements.

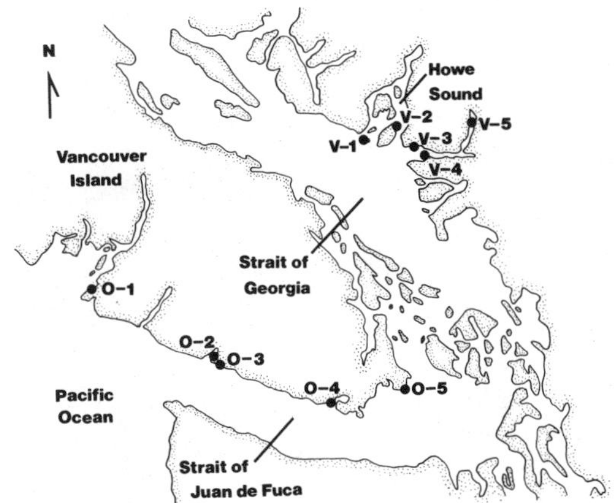

FIGURE 1.—Collection sites for eggs and larvae of tidepool sculpin. Outer-coast sites: O-1, Bamfield; O-2, Port Renfrew; O-3, Botanical Beach; O-4, Sooke; O-5, Victoria. Inland (Vancouver) sites: V-1, Popham Island; V-2, Bowen Island; V-3, West Vancouver; V-4, First Narrows; V-5, Indian Arm. See Table 1 for site and collection information.

For most hatches and plankton tows, samples of 30 larvae were randomly selected for counts, although several hatches yielded fewer than 30 larvae (Table 1). Larvae were measured to the nearest 0.01 mm of notochordal length (tip of snout to tip of urostyle, NL) with dial vernier calipers; the presence or absence of the nape bubble (Figure 2) was scored (1 = present, 0 = absent), and melanophore patterns were quantified. Counts included postanal ventral midline melanophores (PVMM), lateral gut melanophores (LGM), ventral gut melanophore rows (VGR), ventral gut melanophores (VGM), anterior gut melanophores (AGM), and cranial melanophores (CM). Except for CM and VGR counts, all counts were made from the left profile of larvae, including the VGM count.

Population means for all data were estimated with 95% confidence intervals from *t*-values and standard errors. The hypothesis of equal group means was tested with one- or two-way analyses of variance (ANOVA), and statistical differences between group means were identified with Duncan's multiple-range test.

Parental effects were assessed by comparing data for six separate hatches from an inland site near Vancouver, Popham Island (V-1). Geographic trends were compared for pooled data, equal numbers of larvae being drawn randomly from the Popham Island hatches and from four hatches from Port Renfrew (O-2), an open-coast site in the Strait of Juan de Fuca. Larvae from all of the various open-coast and inland sites were compared by group means. Although this study emphasized diagnostic characters of yolk-sac larvae of the tidepool sculpin, developmental

TABLE 1.—Locations and codes for collection sites, collection dates, sample types, and sample sizes (*N*) for collections of tidepool sculpin larvae.

Code	Collection site	Location	Collection date	Sample type[a]	*N*
O-1	Bamfield Inlet	48°50′N, 125°12′W	26 Mar 1986	Hatch	30
O-2	Port Renfrew	48°33′N, 124°24′W	5 Mar 1983	Hatch	8
			5 Mar 1983	Hatch	8
			5 Mar 1983	Hatch	8
			13 Apr 1986	Hatch	30
O-3	Botanical Beach	48°31′N, 124°27′W	5 Mar 1983	Hatch	30
			13 Apr 1986	Hatch	30
O-4	Sooke	48°20′N, 123°44′W	17 May 1987	Hatch	30
			18 May 1987	Hatch	30
			20 May 1987	Hatch	30
O-5	Victoria	48°24′N, 123°19′W	23 Mar 1987	Hatch	30
V-1	SE Popham Island	49°21′N, 123°29′W	3 Apr 1987	Hatch	30
			3 Apr 1987	Hatch	30
			3 Apr 1987	Hatch	30
			3 Apr 1987	Hatch	19
			3 Apr 1987	Hatch	30
			3 Apr 1987	Hatch	30
V-2	NE Bowen Island	49°25′N, 123°18′W	11 May 1987	Tow	30
			11 May 1987	Tow	30
V-3	West Vancouver	49°20′N, 123°12′W	31 Mar 1986	Hatch	30
V-4	First Narrows	49°18′N, 123° 8′W	1 Apr 1986	Hatch	30
V-5	Indian Arm	49°26′N, 122°52′W	22 Apr 1986	Tow	26
			23 Apr 1987	Tow	30

[a]Hatch = larvae from a single egg mass collected from the intertidal zone; tow = larvae of mixed parentage collected by net in a single plankton tow.

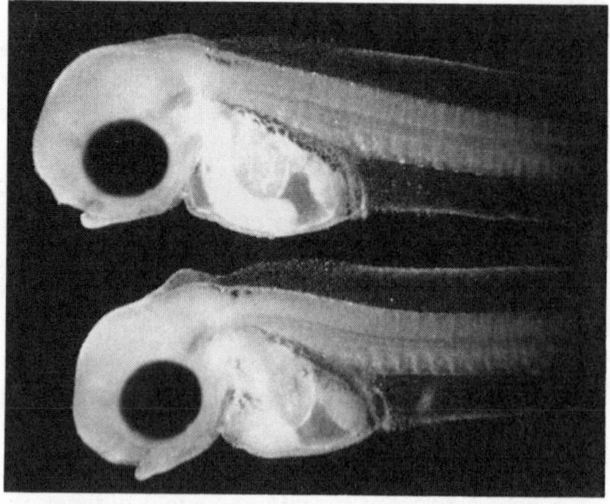

FIGURE 2.—Sibling larval tidepool sculpin, newly hatched. The upper specimen lacks a nape bubble, and the lower one has a nape bubble anterior to the insertion of the dorsal fin fold.

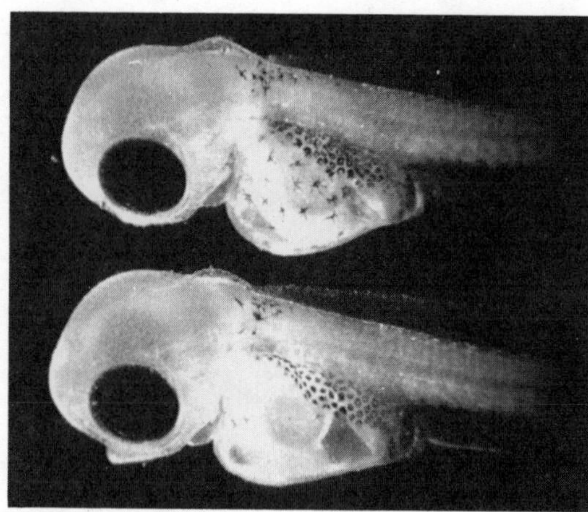

FIGURE 3.—Sibling larval tidepool sculpin, newly hatched, with (above) and without (below) lateral gut melanophores.

changes in characters were verified with a linear correlation matrix and linear regression coefficients. These developmental changes were assessed for 367 larvae, taken in one catch from Bowen Island (V-2), that encompassed all larval stages of this species. From this catch, 30 of the smallest and 30 of the largest larvae were selected randomly and analyzed as separate samples. The sample of smaller larvae from site V-2 was used in the geographic analysis.

Results

The characters investigated differed between hatches within a site. Beyond these parental effects, significant differences in characters existed on a broad geographic basis between open-coast and inland sites. The occurrence of the previously undescribed melanophore patterns (ventral gut, lateral gut, and precocious cranial patterns) varied within a locality. This localized variation was polymorphic, sibling larvae being both with and without patterns such as the lateral gut melanophores (Figure 3).

Parental Effects

Differences between hatches at Popham Island (V-1) were significant ($P < 0.0001$) for all factors considered (Table 2) except cranial melanophores, which were virtually absent. Differences for the presence or absence of the nape bubble

TABLE 2.—Notochordal lengths and melanophore counts for six hatches of tidepool sculpin larvae from the Popham Island site and results of analyses of variance.

	Mean (SD) length or count							
Measure[a]	Hatch 1	Hatch 2	Hatch 3	Hatch 4	Hatch 5	Hatch 6	F	P
NL (mm)	4.42 (0.20)	4.42 (0.13)	4.55 (0.17)	4.18 (0.10)	4.27 (0.20)	4.38 (0.17)	14.60	0.0001
PVMM	36.57 (6.02)	43.53 (4.52)	41.47 (5.81)	37.79 (6.92)	43.60 (5.31)	39.37 (3.98)	8.45	0.0001
LGM	3.93 (3.49)	1.83 (1.51)	0.27 (0.52)	8.84 (5.94)	2.13 (3.54)	0.73 (1.11)	24.21	0.0001
VGM	1.93 (2.03)	0.53 (0.73)	0 (0)	4.05 (3.46)	0.07 (0)	0.27 (0.58)	25.58	0.0001
VGR	1.20 (1.20)	0.73 (0.98)	0 (0)	4.00 (2.98)	0 (0)	0.40 (0.56)	32.62	0.0001
AGM	1.17 (0.95)	2.30 (1.15)	1.63 (0.67)	3.00 (1.37)	2.50 (0.94)	1.77 (1.16)	11.09	0.0001

[a]NL = notochordal length; PVMM = postanal ventral midline melanophores; LGM = lateral gut melanophores; VGM = ventral gut melanophores; VGR = ventral gut (melanophore) rows; AGM = anterior gut melanophores. VGR counts are total counts (ventral view); all other counts are left profile only.

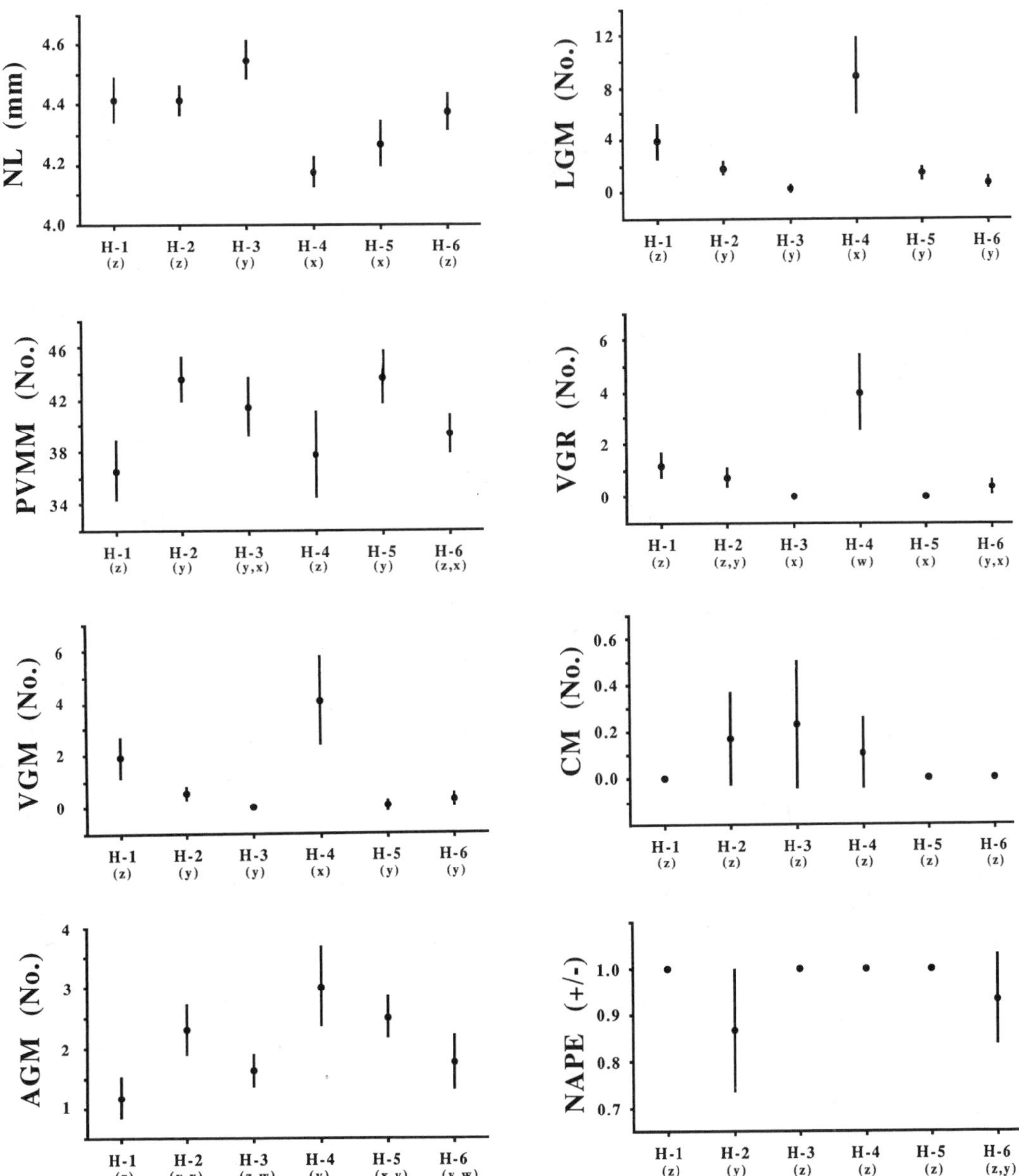

FIGURE 4.—Mean notochord lengths (NL), melanophore counts, and presence or absence of the nape bubble for larval tidepool sculpin from six different hatches (H-1 through H-6, Popham Island site). The vertical lines indicate 95% confidence limits. Means without a letter in common (below the hatch number) are significantly different at $P = 0.05$. Other abbreviations: PVMM = postanal ventral midline melanophores; VGM = ventral gut melanophores; AGM = anterior gut melanophores; LGM = lateral gut melanophores; VGR = ventral gut (melanophore) rows; CM = cranial melanophores.

were only significant at $P = 0.02$. Some agreement existed between characters in their pattern of variation between hatches (Figure 4). In particular, there was linear correlation of ventral with lateral gut melanophores (VGM = $-0.11 + 0.41 \cdot$LGM; $r^2 = 0.67$), as there was for the number of ventral gut melanophores with ventral gut rows (coefficient of determination, $r^2 = 0.81$) and lateral gut melanophores with ventral gut rows ($r^2 = 0.88$). Ventral gut melanophores tended to appear in rows (Figure 5) but, when their densities were low, their staggered array could result in the

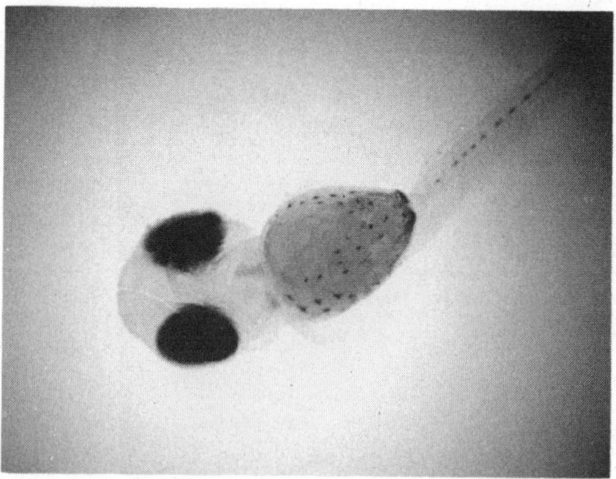

FIGURE 5.—Ventral view of a larval tidepool sculpin, newly hatched, showing presence of ventral gut melanophores arrayed in roughly parallel rows (scored on this specimen as six rows). Note also that lateral gut melanophores are evident on the larva's right side.

leftside VGM count equalling or being lower than the total VGR count (Table 2). These ventral and lateral gut melanophores appeared to be epidermal, on the exterior of the gut area, whereas the anterior gut melanophores and the dorsal gut patch of melanophores (both of which patterns existed on all specimens) were internal. Nape bubbles occurred on most larvae from all six hatches at this site, although the absence of nape bubbles was significantly associated with particular hatches (Figure 4). Although PVMM counts varied between these hatches, all counts were high (means ranging from 36.6 to 43.6) and appeared to involve up to two melanophores for each of the 27 postanal myomeres.

Geographic Effects

A comparison of pooled data from the six Popham Island (V-1) hatches with pooled data from four hatches from an open Pacific coast site, Port Renfrew (O-2), revealed marked differences (Table 3). Specifically, the occurrence of the nape bubble was over 90% at Popham Island versus 40% at Port Renfrew, and the PVMM counts were much lower at Port Renfrew than at Popham Island. The PVMM counts at Port Renfrew involved primarily single melanophores for each postanal myomere. As mentioned above, at Popham Island most postanal myomeres had pairs of ventral melanophores. The occurrence of the LGM pattern was relatively low at Port Renfrew compared to Popham Island, and the correlated VGM pattern was absent. Whether the differences between pooled data for these two geographically distant sites are representative of broad geographic distinction was tested by comparisons of group means for all of the collection sites listed in Table 1 (see also Figure 1). The plots of confidence intervals in Figure 6 show significantly lower PVMM counts at open-coast sites, along with more frequent absences of nape bubbles and VGM and LGM patterns. In contrast, the occurrence of cranial melanophores on newly hatched larvae was more frequent at these open-coast sites.

On a more localized geographic scale, a one-factor ANOVA revealed significant differences in the extent of ventral and lateral gut melanophore patterns on larvae from inland sites. Means (± SD) for the VGM counts in the Howe Sound samples (V-1, V-2) were greater than those for the Burrard Inlet samples (V-3, V-4, V-5) (0.63 ± 1.36 versus 0.14 ± 0.69, respectively; $F = 8.32$; $P = 0.0045$). Similarly, grouped means for LGM counts in Howe Sound averaged 3.42 ± 4.8, versus 1.38 ± 2.2 in Burrard Inlet ($F = 12.37$; $P = 0.0006$). The relatively higher VGM and LGM counts in Howe Sound are comparable to the data for another site bordering the Strait of Georgia at Victoria (O-5). The Victoria site is intermediate geographically; larvae from there had intermediate PVMM counts (Figure 6) and shared features with larvae from both

TABLE 3.—Melanophore counts and presence or absence of a nape bubble for pooled data for six hatches of tidepool sculpin larvae from Popham Island (V-1) versus four hatches from Port Renfrew (O-2) and results of analyses of variance.

	Mean (SD) melanophore count or occurrence of nape bubble			
Measure[a]	Port Renfrew	Popham Island	F	P
PVMM	18.6 (3.2)	39.0 (6.0)	273.9	0.0001
LGM	0.03 (0.2)	2.2 (3.2)	14.4	0.0004
VGM	0 (0)	0.9 (1.5)	9.7	0.0029
Nape	0.4 (0.5)	1.0 (0.2)	34.2	0.0001

[a]PVMM = postanal ventral midline melanophores; LGM = lateral gut melanophores; VGM = ventral gut melanophores; nape = nape bubble present (1) or absent (0).

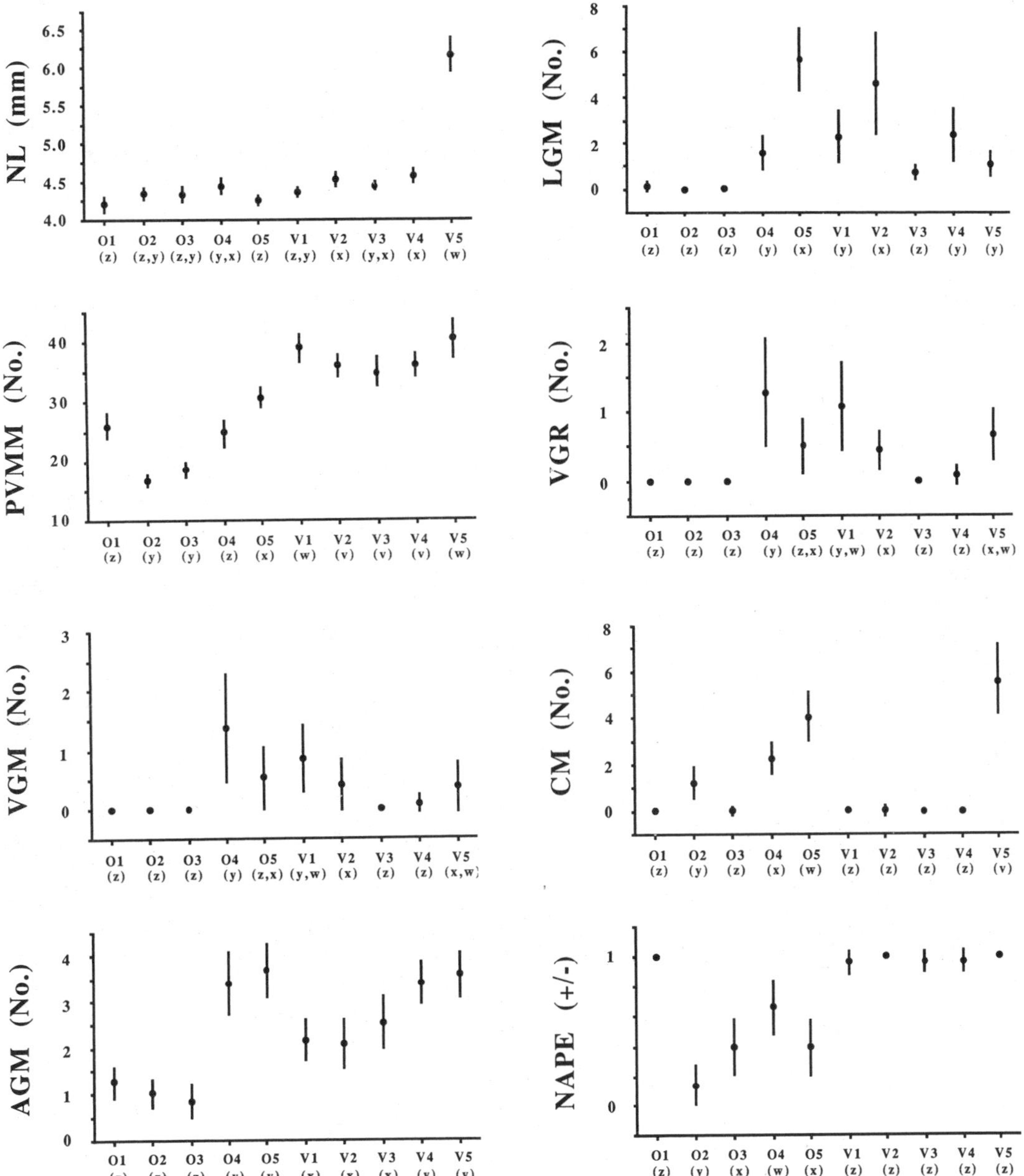

FIGURE 6.—Mean melanophore counts and presence or absence of a nape bubble for tidepool sculpin larvae from outer coast (O-1 to O-5) and inland (V-1 to V-5) sites. Vertical lines are 95% confidence limits. Means without a letter in common (below the site number) are significantly different at $P = 0.05$. Larvae from site V-5 were not in the yolk-sac stage, hence notochord length (NL) and number of cranial melanophores (CM) vary from those of larvae from other sites. Other abbreviations: PVMM = postanal ventral midline melanophores; VGM = ventral gut melanophores; AGM = anterior gut melanophores; LGM = lateral gut melanophores; VGR = ventral gut (melanophore) rows.

the Howe Sound sites (VGM and LGM counts) and the Strait of Juan de Fuca sites (absence of nape bubble, precocial cranial pigment).

Only limited numbers of larvae were caught in tows at the most inland site (V-5), so significantly larger larvae were included in the samples for that site (Figure 6). These larger, older larvae from V-5 were expected to have more developed cranial melanin (Washington 1981), so their data for that character are an artifact of length in Figure 6. The

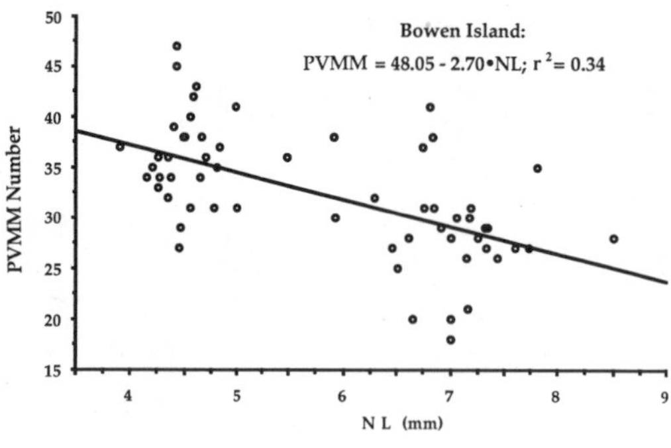

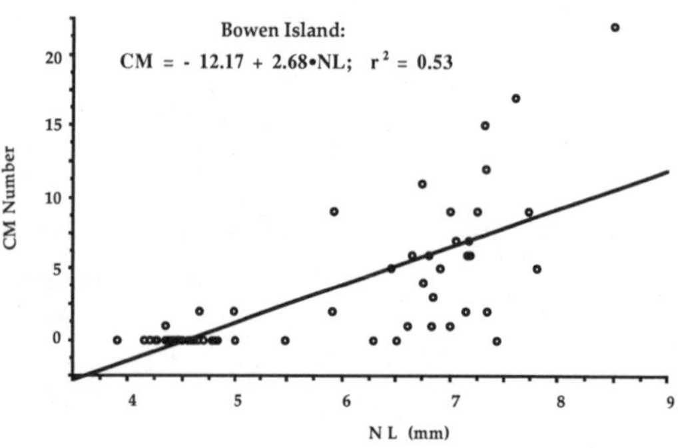

FIGURE 7.—Regressions of numbers of postanal ventral midline melanophores (PVMM, above) and cranial melanophores (CM, below) against notochordal length (NL) for field-caught tidepool sculpin larvae (N = 60) from Bowen Island (site V-2), illustrating coincidental reductions in PVMM counts and increases in CM counts with growth in length.

two size-groups of field-caught larvae from Bowen Island (V-2) provided data for analysis of such developmental changes. Along with the expected disappearance (Washington 1981) of the nape bubble during flexion, PVMM counts decreased linearly 2.7 counts for each 1-mm increase in length (Figure 7), and cranial melanophore counts increased at the same rate with increasing length (Figure 7). Melanophores rose internally along myosepta among more anterior PVMMs of postflexion larvae, then corresponding gaps appeared in larger larvae. It should be noted from the data in Figure 7, however, that the largest larvae from this inland site averaged fewer than 10 cranial melanophores, whereas some of the newly hatched larvae from Victoria had that many.

Discussion

Natural selection acts on every life stage, independently of other stages, and we must expect variability in larvae just as we do in adults, although not necessarily the same types of variability. Indeed, Strauss and Fuiman (1985) found, in a morphometric analysis of larvae versus adults of five Pacific sculpin species, that larvae tend to be more variable within species than adults. The tidepool sculpin represents a diverse fauna of rocky nearshore fishes that avoid the dispersive effects of planktonic drift by schooling as larvae very close to shore (Marliave 1986), so the possibility exists for localized genetic divergence of populations. The physical conditions experienced by tidepool sculpin embryos and larvae vary considerably over the species' range, from relatively high-salinity seawater and considerable surge and wave action along open-coast shorelines to lower salinities and protection from ocean swells inland. For example, during the winter–spring period of this species' larval stage, average salinity varies from 15 to 29‰ in the Strait of Georgia (LeBlond 1983), and Indian Arm (V-5) surface salinities can drop to 7‰ during heavy runoff (personal observations).

Whether salinity is important or not, the increased melanic pigmentation of yolk-sac tidepool sculpin larvae in these inland waters is matched by increased melanism of unrelated larval fishes. In the Strait of Georgia, larval Pacific hake *Merluccius productus* (Gadidae) have black pectoral fins (personal observation) not described from elsewhere (see Ahlstrom and Counts 1955) and preflexion larvae of the starry flounder *Platichthys stellatus* (Pleuronectidae) have similarly undocumented melanophores along preanal and dorsal fin fold margins (personal observation).

The increase in postanal ventral midline melanophores appears to be clinal on a broad geographic basis; they are fewest on larvae in the California and Oregon area (11–20: Stein 1973; Washington 1981), increase through the Strait of Juan de Fuca (12–41, present study), and are most numerous in the Strait of Georgia region (22–53, present study). Washington (1981) described PVMMs starting several myomeres posterior to the anus and occurring on every second or third myomere in the anterior portion of the pigment row. Progressive increases in PVMM counts appear to involve first melanophores that occur more regularly on every myomere, then melanophores that occur further anteriorly toward the anus, and then two melanophores each on increasing proportions of myomeres. This progression itself occurs clinally.

The absence of a nape bubble, precocial development in cranial melanophores, and occurrence

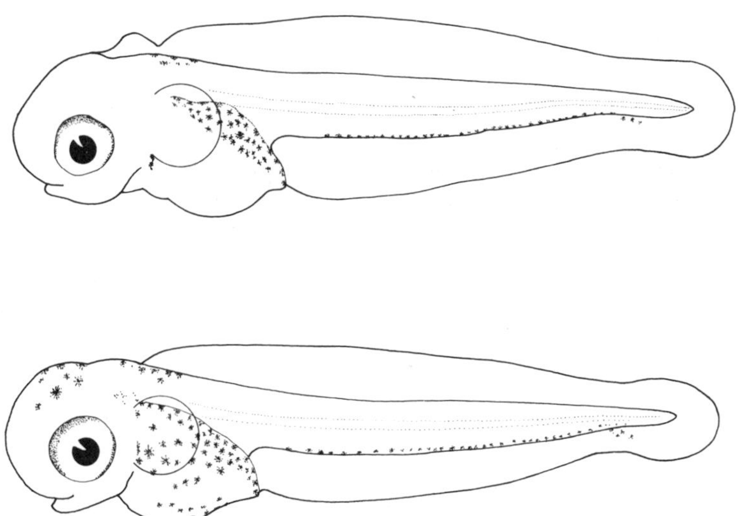

FIGURE 8.—Yolk-sac larvae (both 4.5 mm in notochord length) of the tidepool sculpin from British Columbia (present study, myomeres not illustrated). The upper larva conforms to published descriptions (Stein 1973; Washington 1981) except for a higher postanal ventral midline melanophore count. The lower larva cannot be reconciled with published literature owing to the presence of ventral gut melanophore rows, lateral gut melanophores, and cranial melanophores, together with the absence of a nape bubble.

of ventral and lateral gut melanophore patterns all vary on a more localized scale and show more pronounced variability between hatches within sites. There are localized regional trends, however, such as the absence of the nape bubble in the open-coast area of the Strait of Juan de Fuca. Existing literature indicates that this trend does not extend to outer coastal areas generally, because the California and Oregon specimens all had nape bubbles (Stein 1973; Washington 1981).

Existing descriptions of tidepool sculpin larvae from California (Stein 1973) and Oregon and Washington (Washington 1981) are inadequate for identification of some tidepool sculpin larvae from British Columbia (Figure 8). The use of reared larvae from single hatches or samples from a limited portion of a species' range introduces the risk of a narrow description in terms of pigment variation. With larvae of nearshore marine species or other fishes ranging across variable environments, it may be especially important to obtain geographically diverse samples or hatches. For description of a laboratory-reared developmental series, diverse parental sources might be essential for an accurate indication of pigment variation, whether or not diverse geographic sources are feasible.

Claims that laboratory rearing results in larvae with abnormal pigmentation have not been borne out in the literature. Watson (1982) described larvae of a sciaenid species having variable pigment and stated that reared larvae were often more heavily pigmented than field specimens. Hunter (1984) proposed that heavier pigment in reared larvae is due to the expansion, not added numbers, of pigment cells. Indeed, Mansfield and Mansfield (1982) demonstrated a significant effect of background color on melanophore expansion in reared larvae of the spotfin shiner *Notropis spilopterus*. Reared larvae tend to be killed by fixation whereas field-caught larvae often die in the net prior to fixation and always experience relatively prolonged stress prior to introduction of preservatives. Theilacker (1980) and Hay (1981) demonstrated that death in nets prior to fixation caused greater shrinkage and alteration of morphometric characters than preservation itself. A procedure of holding reared larvae (cottid or stichaeid) 10 min in a submerged dip net prior to killing them will similarly cause marked contraction of melanophores (personal observations). Hunter (1984) also offered a conjecture that earlier occurrence of pigment development may occur in reared larvae. Such a variation is likely due to limited genetic sources and greater pigment expansion, and not to abnormality resulting from culture conditions. The present study documents precocial development of cranial pigment in tidepool sculpins from one particular region and in particular hatches from that region. It therefore appears less likely that laboratory rearing produces abnormalities in pigmentation than that limited sources and a different manner of killing account for the differences from field-caught specimens.

Acknowledgments

I thank Hector Leon for statistical testing, Michelle Kerr and Renee Giroux for laboratory assistance, and Ernest Cooper for field assistance. I also thank my director, Murray Newman, and my patron, Rudy North, for providing support for this project.

References

Ahlstrom, E. H., and R. C. Counts. 1955. Eggs and larvae of the Pacific hake *Merluccius productus*. U.S. Fish and Wildlife Service Fishery Bulletin, 56: 295–329.

Bosley, T. R., & J. V. Conner. 1984. Geographic and temporal variation in numbers of myomeres in fish larvae from the lower Mississippi River. Transactions of the American Fisheries Society 11:238–242.

Hay, D. E. 1981. Effects of capture and fixation on gut contents and body size of Pacific herring larvae. Rapports et Procès-Verbaux des Réunions, Conseil International pour l'Exploration de la Mer 178:395–400.

Hunter, J. R. 1984. Synopsis of culture methods for marine fish larvae. American Society of Ichthyologists and Herpetologists Special Publication 1:24–27.

Kendall, A. W., Jr., E. H. Ahlstrom, and H. G. Moser. 1984. Early life history stages of fishes and their characters. American Society of Ichthyologists and Herpetologists Special Publication 1:11–22.

LeBlond, P. H. 1983. The Strait of Georgia: functional anatomy of a coastal sea. Canadian Journal of Fisheries and Aquatic Sciences 40:1033–1063.

Mansfield, P. J., and A. H. Mansfield. 1982. Influence of background color and intensity of illumination on melanophore expansion in larval fish. Pages 60–62 *in* C. F. Bryan, J. V. Conner, and F. M. Truesdale, editors. The fifth annual larval fish conference. Louisiana State University, Louisiana Cooperative Fishery Research Unit, Baton Rouge.

Marliave, J. B. 1986. Lack of planktonic dispersal of rocky intertidal fish larvae. Transactions of the American Fisheries Society 115:149–154.

Powles, H., and D. F. Markle. 1984. Identification of larvae. American Society of Ichthyologists and Herpetologists Special Publication 1:31–33.

Stein, R. 1973. Description of laboratory reared larvae of *Oligocottus maculosus* Girard (Pisces: Cottidae). Copeia 1973:373–377.

Strauss, R. E., and L. A. Fuiman. 1985. Quantitative comparisons of body form and allometry in larval and adult Pacific sculpins (Teleostei: Cottidae). Canadian Journal of Zoology 63:1582–1589.

Theilacker, G. H. 1980. Changes in body measurements of larval northern anchovy *Engraulis mordax* and other fishes due to handling and preservation. U.S. National Marine Fisheries Service Fishery Bulletin 78:685–692.

Washington, B. B. 1981. Identification and systematics of larvae of *Artedius*, *Clinocottus*, and *Oligocottus* (Scorpaeniformes: Cottidae). Master's thesis. Oregon State University, Corvallis.

Watson, W. 1982. Development of eggs and larvae of the white croaker, *Genyonemus lineatus* Ayres (Pisces: Sciaenidae), off the southern California coast. U.S. National Marine Fisheries Service Fishery Bulletin 80:403–417.

American Fisheries Society Symposium 5:89–95, 1988

Marking Cyprinid Larvae with Tetracycline[1]

ROBERT T. MUTH

Larval Fish Laboratory, Department of Fishery and Wildlife Biology
Colorado State University, Fort Collins, Colorado 80523, USA

THOMAS P. NESLER

Wildlife Research Unit, Colorado Division of Wildlife
Fort Collins, Colorado 80526, USA

ANTHONY F. WASOWICZ[2]

Department of Fishery and Wildlife Biology, Colorado State University
Fort Collins, Colorado 80523, USA

Abstract.—Laboratory experiments were conducted on the use of tetracycline hydrochloride (TC) for mass marking recently hatched Colorado squawfish *Ptychocheilus lucius*. Protolarvae were immersed in tris-buffered solutions of 200, 350, or 500 mg TC/L of distilled water for 4, 12, or 36 h and then reared for up to 77 d. Overall survival was lowest (0–81%) for the 36-h and 500-mg/L treatments. However, fish growth and development were not affected by treatment. Larvae preserved within 15 d of treatment exhibited externally visible fluorescence when scanned with ultraviolet (UV) light. Otoliths extracted from larvae that were preserved in 95% ethanol were in good condition, but those from larvae preserved in formalin solutions buffered to near neutral pH with phosphate degraded during storage. The otoliths of treated larvae exhibited fluorescent marks when examined by UV-light microscopy, and these marks were retained throughout the experiment with no apparent decrease in intensity. Whole-body immersion in 350 mg TC/L for 4–12 h was found to be optimal for best mark intensity and fish survival. In another experiment, protolarvae of fathead minnows *Pimephales promelas* were immersed in a tris-buffered solution of 350 mg TC/L for 4 h and then exposed to various intensities of artificial white light in the 340–650-nm range (includes near-UV spectrum) for 12 h/d for 7 d. Exposure to light had no effect on the presence or intensity of fluorescent marks in the otoliths. Our results suggest that the TC-marking technique can be used in field mark–recapture studies of Colorado squawfish larvae.

Marking individual fish is an effective means for obtaining various kinds of information on fish species and populations (Wydoski and Emery 1983). Data from mark–recapture studies often provide solutions to questions left unanswered by basic survey sampling. Tsukamoto (1985) observed that information on the early-life behavior and environmental requisites of fish is critical to understanding the dynamics of fish populations, and he noted that effective methods of marking fish early-life stages for mark–recapture studies need to be developed. Marking techniques used successfully for the early-life stages of some fish species might require modification or be unsuitable for use on other fish species.

The objective of this work was to develop or adapt a mass-marking technique for larvae of the federally endangered Colorado squawfish *Ptychocheilus lucius* for proposed mark–recapture studies in the Yampa and Green rivers in Colorado and Utah. This technique would facilitate investigation of dispersal patterns of larvae, habitat utilization by age-0 fish, relationships between larval fish abundance and recruitment, and causes of early-life mortality. For such studies, an optimal mark must be (1) applicable to late embryos or recently hatched, yolk-bearing larvae or both, (2) suitable for mass marking in the field, (3) innocuous, that is, must not affect the normal life of the fish, (4) detectable for at least 8 weeks, and (5) relatively easy and inexpensive to apply and detect.

Techniques for marking very small fish were reviewed by Laird and Scott (1978), Hettler (1984), Brothers (1985), and Tsukamoto (1985). These techniques included the use of dyes or stains, fluorescent antibiotics, trace element or nonradioactive isotopes, and growth inhibitors or environmental manipulations (e.g., of tempera-

[1]Contribution 35, Colorado State University Larval Fish Laboratory.

[2]Present address: Utah Cooperative Fish and Wildlife Unit, Utah State University, Logan, Utah 84321, USA.

ture or light) to alter growth patterns on bony structures. Many marking studies of fish early-life stages have been primarily concerned with age or growth determinations. Few successful methods for marking fish embryos or early larvae, specifically for application in field investigation, have been described (e.g., Muncy and D'Silva 1981; Brothers 1985; Tsukamoto 1985). Brothers (1985) and Tsukamoto (1985) considered otoliths ideal for marking because they are the first permanent calcified structures present in the earliest life history stages of fish and are, effectively, biological internal tags. We concluded that tetracycline marking of otoliths would most likely meet our requirements for an optimal mark. Tetracycline compounds are well-known markers for calcified structures, and they fluoresce yellow under ultraviolet (UV) light (Milch et al. 1957; Weber and Ridgway 1962, 1967; Choate 1964). They have been used to mark larvae of several fish species, for example, pinfish *Lagodon rhomboides* and spot *Leiostomus xanthurus* (Hettler 1984) and ayu *Plecoglossus altivelis* (Tsukamoto 1985), but their application to cypriniform larvae has not been reported.

Methods

Approximately 1,500 4-d-old (posthatching) Colorado squawfish protolarvae (larvae prior to development of median fin rays, sensu Snyder 1981) were obtained from the Dexter (New Mexico) National Fish Hatchery in June 1986. The larvae were placed in a flow-through holding trough that received 20°C well water; 20–22°C water was used by Hamman (1986) for hatchery culture of Colorado squawfish embryos and larvae. Most larvae were swim-up, still had 50–70% of their yolk, and measured 7.5–8.0 mm total length (TL). The sagitta and lapillus otoliths were both present. After a 24-h acclimation period, dead, damaged, or atypically behaving larvae were removed. The remaining larvae were divided into ten 140-specimen experimental groups.

Nine experimental groups were treated by placing the larvae in 1-L glass beakers containing tetracycline hydrochloride (TC) solutions of 200, 350, or 500 mg TC/L aerated distilled water. Distilled water was used as the diluent because TC binds to calcium, and hard water, which contains calcium ions, might hinder the uptake of TC by the otoliths. Each TC test solution was adjusted to pH 6.8–7.0 with tris buffer (Hettler 1984). The pH prior to buffering was about 3.6. Exposure times for each concentration were 4, 12, and 36 h. During their exposure, the treatment groups were aerated and incubated in a covered water bath at 20°C, but they were not fed. The 10th experimental group was maintained as a control.

After each prescribed exposure, the treated larvae were removed from the test solutions, and the numbers of dead larvae were recorded. Surviving larvae were placed in aerated well water in 3.8-L rearing jars and incubated in a water bath at 20°C. Control larvae were taken directly from the holding trough and placed in a rearing jar. Rearing jars were partially shaded with sheets of opaque plastic. The photoperiod was about 9-h light: 15-h dark; light intensity was 100 lx. Larvae were fed twice daily with live artemia *Artemia* sp. nauplii and TetraMin Fry Diet. After about 4 weeks, TetraMin Staple Food was included in the diet (Muth et al. 1985). Before each feeding, the rearing jars were cleaned, and the numbers of dead larvae were recorded. Larvae were reared for up to 77 d after treatment.

Ten living larvae from each experimental group were preserved and measured (TL) immediately after treatment and at weekly intervals thereafter. Two different preservation fluids were tested, and their effects on otolith structure and on TC-produced marks were compared. At each time interval, five larvae were fixed and preserved in 95% ethanol (pH = 8.0), and five were fixed and preserved in formalin solutions buffered to pH 6.8 with phosphate (Markle 1984). Larvae were fixed in 10% buffered formalin for 24 h, then transferred to 3% buffered formalin for storage. Samples were stored in the dark until they were examined 3–4 months after treatment.

Two whole larvae from each sample lot were examined with incident UV light under a dissecting microscope for the presence of external fluorescent marks. Sagittae and lapilli were extracted from larvae by procedures similar to those described by Brothers (1987) for embryonic and larval fish. Otoliths were mounted in glycerin on glass slides and examined with incident UV light under a compound microscope for the presence and intensity of fluorescent marks. The average time spent on the extraction and mounting of otoliths was about 10 min/specimen. The intensity of the mark was determined for each of the larvae that was examined, and it was ranked by a procedure similar to that used by Tsukamoto (1985). The ranking categories that we designated were absent, faint, lucid, or bright, corresponding to values of 0, 1, 2, or 3, respectively. Individual values were summed within each sample lot, and each total was used to compare the intensities of the marks among the sample lots. For example, the maximum summed mark-intensity value for a sample lot of five specimens would be 15.

A second experiment was conducted to examine the stability of TC deposited in otoliths of fish larvae that were then exposed to white light. About 300 7-d-old (posthatching) fathead minnow *Pimephales promelas* protolarvae were immersed in an aerated, tris-buffered solution of 350 mg TC/L for 4 h. Larvae were 5.5–6.5 mm TL, and lapilli and sagittae were present. After immersion, five larvae were preserved in 95% ethanol. The remaining larvae were divided equally into six experimental groups and placed in 38-L aquaria containing aerated, dechlorinated tap water at 22°C. Then, the larvae were exposed to white light at replicated treatment intensities of 10, 300, or 1,000 lx (measured at the water surface) for 12 h/d for 7 d. The two lower light intensities were achieved by covering the aquaria with sheets of opaque plastic or fine-mesh screening. Illumination was provided by paired General Electric cool white 40-W fluorescent tubes positioned 20 cm above the aquaria. According to the manufacturer, these tubes emit light in the 340–650-nm range. Weber and Ridgway (1967) stated that tetracycline molecules that were deposited in fish bone were excited to a fluorescent state by absorbing UV energy at 360 nm. Larvae were fed twice daily with live artemia nauplii. Larvae in each light-intensity treatment were sampled once daily and preserved in 95% ethanol. Then, otoliths were extracted, mounted, and examined in the same manner as for Colorado squawfish larvae.

Results

Mortality of Colorado squawfish larvae was high for all 36-h-exposure treatments; 71–100% of the fish in each group died. In the 500-mg/L treatment for 4- and 12-h, mortality was moderate (16 and 24% of each group, respectively), then it stabilized nearly to zero during rearing. For the remaining treatments, survival was high during exposure to TC and during rearing. Overall survival (90–92%) of each treatment group was higher than that of the control group (79%). Fish growth, measured in increments of total length, was similar for all experimental groups. Exposure to TC had no obvious effect on fish development or behavior.

Fluorescence was visible externally when whole treated Colorado squawfish larvae were scanned with UV light. The entire bodies of larvae

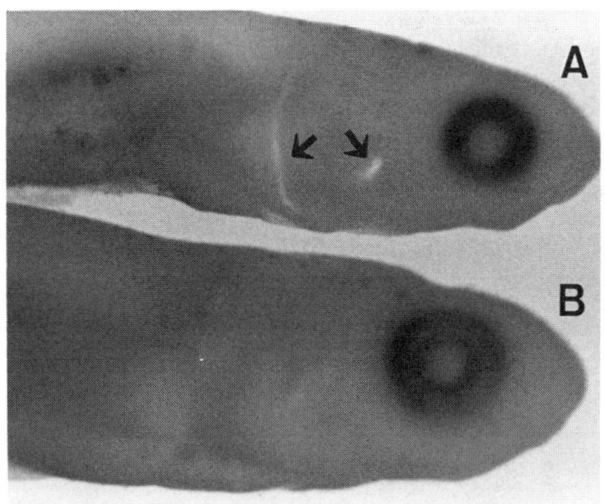

FIGURE 1.—**A.** Externally visible fluorescent marks (arrows) on preopercal and subopercal bones of a Colorado squawfish larva (10.7 mm total length, TL) scanned with ultraviolet (UV) light 14 d after whole-body immersion in a 350-mg tetracycline hydrochloride/L solution for 4 h. **B.** Untreated control larva (10.5 mm TL) scanned with UV light.

from all treatments that had been preserved in ethanol or buffered formalin within 15 d of exposure to TC fluoresced pale yellow. Fluorescence was most evident in the yolk and gut regions. Larvae that had been preserved in ethanol revealed distinct yellow fluorescent marks on the developing dorsoanterior tip of the preopercal bone and posterior edge of the subopercal bone (Figure 1). Fluorescent marks on these bones were not present in larvae that had been preserved in buffered formalin. Larvae that had been preserved 2 weeks or more after treatment revealed externally visible fluorescence that was indistinguishable from reflections caused by incident illumination.

The otoliths extracted from larvae that had been preserved in ethanol appeared structurally intact. However, the otoliths of larvae that had been preserved in buffered formalin had either disintegrated during storage or degenerated to thin, transparent disks that crumbled upon handling. These otoliths were unsuitable for microscopic examination. Asteriscus otoliths were formed in larval Colorado squawfish during the third week of rearing, and they were not examined for fluorescent marks.

Readily detectable marks were visible in otoliths of all TC-treated Colorado squawfish larvae examined by UV-light microscopy (Table 1 and

TABLE 1.—Relative intensity of fluorescent marks in otoliths from Colorado squawfish larvae that were immersed in tetracycline hydrochloride (TC) solutions. Mark intensities were determined for five larvae from each treatment and sampling time (days after treatment) and were ranked 0–3 (absent–bright). Values were summed within each sample lot to produce a batch mark-intensity value. All fish immersed in TC solutions of 350 or 500 mg/L for 36 h died during treatment.

Days after treatment	Mark intensity by TC concentration and exposure time						
	200 mg/L			350 mg/L		500 mg/L	
	4 h	12 h	36 h	4 h	12 h	4 h	12 h
0	5	9	10	10	14	10	14
7	5	10	12	12	15	13	14
28	5	8	12	15	14	14	15
56	5	10		13	13	15	14
77	5	9		13	14	14	14

Figure 2). Marks appeared as yellow fluorescent bands against a dark-green background and were retained throughout the 11-week experiment with no apparent decrease in intensity over time. Fluorescent marks in sagittae became slightly more diffuse as otoliths changed shape with growth; however, mark intensity was not affected (Figure 2). Overall, mark intensity was higher for the 350- and 500-mg/L treatments than for the 200-mg/L treatments, but mark intensity increased with exposure time at all treatment concentrations. In the second experiment, otoliths of all fathead minnow larvae examined by UV-light microscopy had fluorescent marks similar to those observed in otoliths of Colorado squawfish larvae. Mark intensity was high for all three light-intensity exposures throughout the 7-d experiment.

Discussion

Our results demonstrated that otoliths of protolarval Colorado squawfish can be successfully marked by whole-body immersion in TC solutions. Tetracycline deposited in otoliths of larval fish has been successfully used as a mark for several species (e.g., Hettler 1984; Tsukamoto 1985), but, to our knowledge, this demonstration is the first documented use for marking the early larvae of cyprinids. The technique is relatively easy to use, reasonable in cost, and suitable for the mass marking of fish in the field; it offers 100% marking success and at lower concentrations and exposure times has no adverse effect on fish mortality, growth, or development. For the best survival and mark intensity, larvae should be immersed in a solution of 350 mg TC/L for 4–12 h. Hettler (1984) and Tsukamoto (1985) reported similar results. Tsukamoto (1985) observed that larval ayu survival decreased as exposure time in 500 mg TC/L increased. This relationship was not noted at lower TC concentrations. The lowest ayu survival occurred in TC concentrations of 500 mg/L or greater for all exposure times tested. Tsukamoto also reported that TC treatment had no affect on ayu growth, and he recommended whole-body immersion in 200–300 mg TC/L for 3–24 h for otolith marking. Hettler (1984) found that the percentage of larval spots with otoliths that were successfully marked by oxytetracycline hydrochloride (OTC, a compound closely related to TC) increased as exposure time was increased, and it reached 100% in OTC concentrations of 250 and 500 mg/L at exposures of 120 and 60 min, respectively.

Mark Retention

The knowledge of mark-retention time is necessary for mark–recapture studies. In our laboratory experiments, TC deposits in otoliths of larval Colorado squawfish were retained in detectable amounts at least 77 d after treatment in all of the fish that were examined. Larval ayu that were treated with 100 mg TC/L for 7 h and then reared in a greenhouse retained fluorescent marks in their otoliths for at least 164 d (Tsukamoto 1985). Several authors (Choate 1964; Trojnar 1973; Laird and Scott 1978; Hettler 1984; Brothers 1985) reported that tetracycline-produced marks are labile in light because sunlight deactivates the fluorescent properties of tetracycline compounds. However, they provided no confirmatory evidence that marks in internal bony structures would be lost or degraded under natural conditions. Deactivation of tetracycline is presumably caused by molecular excitation and subsequent decay through absorption of solar energy in the UV spectrum (particularly at 360 nm, Weber and Ridgway 1967). Excited molecules of fluorescent substances are usually unstable and have a mean lifetime of only about 10^{-8}–10^{-9} s (Lehninger 1975). If this were true, the reliability of tetracycline marking of larval fish for use in field mark–recapture studies would be suspect. However, this presumption is questionable, particularly for TC deposited in the otoliths of fish released into natural waters. In our experiment using fathead minnow larvae treated with TC, exposure of larvae to artificial white light (in the 340–650 nm range) at surface light intensities as high as 1,000 lx during rearing in pure tap water had no effect on the presence or intensity of fluorescent marks in otoliths. Water absorbs UV

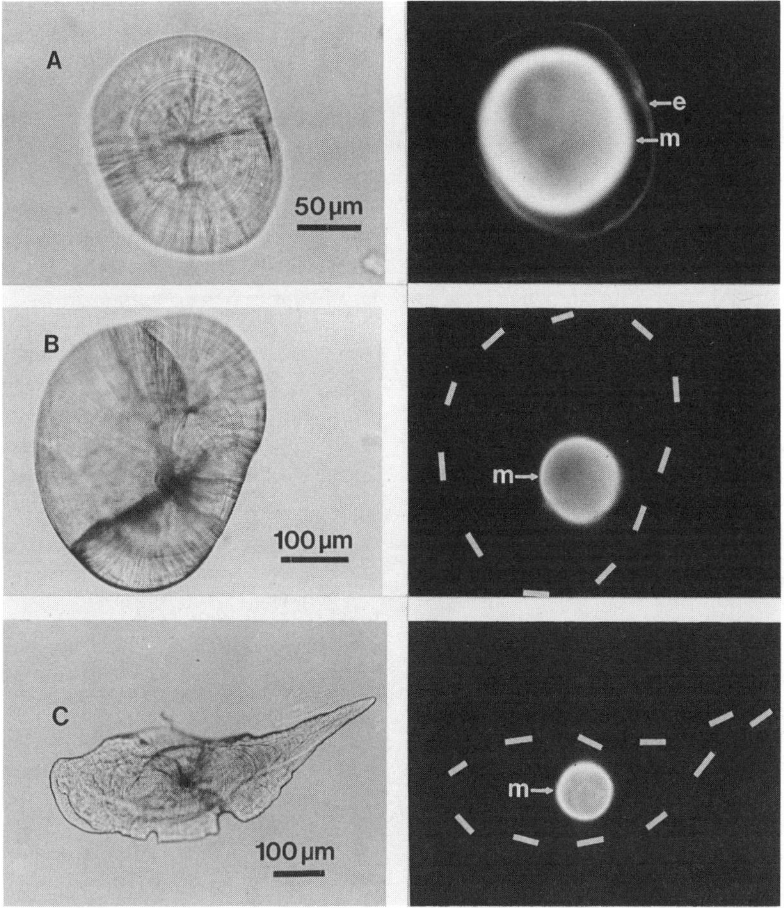

FIGURE 2.—Microphotographs of unground otoliths under white (left) and ultraviolet (UV) light (right). Otoliths from Colorado squawfish larvae immersed in a 350-mg tetracycline hydrochloride/L solution for 4 h. **A.** Sagitta from a 9.1-mm total length (TL) larva, 7 d after treatment. m = fluorescent mark; e = edge of otolith under UV light. **B.** Lapillus and **C.** Sagitta from a 18.2-mm TL larva, 56 d after treatment. m = fluorescent mark; edge of otoliths under UV light indicated by dashed line.

light, and its absorption capacity increases with increased amounts of dissolved solids and organic matter (Koller 1965). Under natural conditions, the small amount of solar UV radiation that penetrates the atmosphere and strikes the earth is attenuated in the surface layer of natural waters (Nicol 1974; Golterman 1975; Reid and Wood 1976; Ali and Klyne 1985; Kirk 1986). Also, the skull and skin of fish should directly protect internal structures from UV radiation (Koller 1965). Skin pigments, particularly melanin on the dorsal surface (including the head and nape), afford protection from light. Melanin helps protect underlying tissues from UV radiation by scattering rays, absorbing damaging radiation (less than 400 nm), and dissipating absorbed energy as heat (Pathak and Stratton 1969). Finally, the successful use of tetracycline-marked fish has been reported in field mark–recapture studies. Blackler (1974) indicated that tetracycline marks may persist for at least several years in internal structures such as bones and otoliths. Pacific salmon *Oncorhynchus* spp., fed pelleted food that contained tetracycline and then released into the wild, retained fluorescent marks in internal bone for at least 3.5 years after treatment (Weber and Ridgway 1967). Kokanee *Oncorhynchus nerka*, fed OTC-bearing food as larvae and early juveniles and then released into Lake Granby, Colorado, retained fluorescent marks in caudal vertebrae for 5–6 years (W. J. Wiltzius, Colorado Division of Wildlife, personal communication).

The utility of external fluorescence produced by exposure to tetracycline as a mark for young fish appears limited by its relatively short time of retention and detectability after treatment. Brothers (1985) reported observations similar to ours for larvae and early juveniles of lake trout *Salvelinus namaycush* exposed to TC and suggested that the presence of external fluorescence after treatment is a good indicator of otolith marking success. Hettler (1984) noted that the head and fins of larval spot and pinfish preserved 8 d after immersion in OTC solutions fluoresced yellow under longwave UV light. In mark–recapture studies of short duration in which larvae are released and sampled soon after marking, external fluorescence might permit identification of marked fish without sacrificing them.

Specimen Preservation

In this study, the otoliths removed from Colorado squawfish larvae that were fixed and preserved in 95% ethanol (as recommended by Brothers 1987) were in excellent condition, but the larvae themselves were shrunken and deformed due to dehydration. Conversely, larvae that were fixed and preserved in formalin solutions buffered to pH 6.8 were in good morphological condition, but their otoliths were either disintegrated or were significantly degraded and useless. The specific cause for loss or degradation of the otoliths is not known. Because the skeletal features of larvae preserved in formalin solutions buffered nearly to neutrality were successfully retained, we considered that 10% formalin fixative and 3% formalin preservative, both made with distilled water and buffered to pH 6.8 with phosphate (Markle 1984), would be adequate for the preservation of otoliths. Perhaps alkaline formalin solutions (about pH 8.0) are required as suggested by Steedman (1976) and McMahon and Tash (1979), or water content in the solutions (greater than 96%) was too high (Steedman 1976), or both. Perhaps, calcium ion concentration in the preservative should have been near saturation rather than almost absent (as in solutions made with distilled water). For the present, although we prefer and recommend formalin solutions buffered nearly to neutrality for most other purposes, we can only follow other researchers and recommend that specimens be preserved in concentrated alcohol (e.g., 95% ethanol) or frozen for otolith analysis.

Acknowledgments

This study was funded under Federal Aid in Fish Restoration matching grant 02-01-035 SE-3 administrated through the Wildlife Research Unit, Colorado Division of Wildlife. We thank J. Hamill (U.S. Fish and Wildlife Service, Denver, Colorado) for arranging the required endangered species permits, R. Hamman (Dexter National Fish Hatchery, Dexter, New Mexico) for providing Colorado squawfish larvae, and D. Keefe (Environmental Research and Technology, Incorporated, Fort Collins, Colorado) for providing fathead minnow larvae. R. Behnke, C. Carlson, and D. Snyder reviewed earlier drafts.

References

Ali, M. A., and M. A. Klyne. 1985. Vision in vertebrates. Plenum, New York.

Blackler, R. W. 1974. Recent advances in otolith studies. Pages 67–90 *in* F. R. Harden Jones, editor. Sea fisheries research. Wiley, New York.

Brothers, E. B. 1985. Otolith marking techniques for the early life history stages of lake trout. Great Lakes Fishery Commission, Research Completion Report, Ann Arbor, Michigan.

Brothers, E. B. 1987. Methodological approaches to the examination of otoliths in aging studies. Pages 319–330 *in* R. C. Summerfelt and G. E. Hall, editors. Age and growth of fish. Iowa State University Press, Ames.

Choate, J. 1964. Use of tetracycline drugs to mark advanced fry and fingerling brook trout (*Salvelinus fontinalis*). Transactions of the American Fisheries Society 93:309–311.

Golterman, H. L. 1975. Physiological limnology. Elsevier Scientific, Amsterdam.

Hamman, R. L. 1986. Induced spawning of hatchery-reared Colorado squawfish. Progressive Fish-Culturist 48:72–74.

Hettler, W. F. 1984. Marking otoliths by immersion of marine fish larvae in tetracycline. Transactions of the American Fisheries Society 113:370–373.

Kirk, J. T. 1986. Light and photosynthesis in aquatic ecosystems. Cambridge University Press, Cambridge, England.

Koller, L. R. 1965. Ultraviolet radiation. Wiley, New York.

Laird, L. M., and B. Scott. 1978. Marking and tagging. IBP (International Biological Programme) Handbook 3:84–100.

Lehninger, A. L. 1975. Biochemistry. Worth, New York.

Markle, D. F. 1984. Phosphate buffered formalin for long term preservation of formalin fixed ichthyoplankton. Copeia 1984:525–527.

McMahon, T. E., and J. C. Tash. 1979. Effects of formalin (buffered and unbuffered) and hydrochloric acid on fish otoliths. Copeia 1979:155–156.

Milch, R. A., D. P. Rall, and J. E. Tobie. 1957. Bone localization of the tetracyclines. Journal of the National Cancer Institute 19:87–93.

Muncy, R. J., and A. P. D'Silva. 1981. Marking walleye eggs and fry. Transactions of the American Fisheries Society 110:300–305.

Muth, R. T., C. M. Haynes, and C. A. Carlson. 1985. Culture of roundtail chub, *Gila robusta robusta* (Cyprinidae), through the larval period. Southwestern Naturalist 30:152–154.

Nicol, J. A. 1974. Studies on the eyes of fishes: structure and ultrastructure. Pages 579–607 *in* M. A. Ali, editor. Vision in fishes. Plenum, New York.

Pathak, M. A., and K. Stratton. 1969. Effects of ultraviolet visible radiation and the production of free radicals in skin. Pages 207–222 *in* F. Urbach, editor. The biologic effects of ultraviolet radiation. Pergamon Press, Oxford, England.

Reid, G. K., and R. D. Wood. 1976. Ecology of inland waters and estuaries. Van Nostrand, New York.

Snyder, D. E. 1981. Contributions to a guide to the cypriniform fish larvae of the upper Colorado River system in Colorado. U.S. Bureau of Land Management, Biological Sciences Series 3, Denver, Colorado.

Steedman, H. F. 1976. Cell products: calcium salts. Monographs on Oceanographic Methodology 4: 209–221.

Trojnar, J. R. 1973. Marking rainbow trout fry with tetracycline. Progressive Fish-Culturist 35:52–54.

Tsukamoto, K. 1985. Mass-marking of ayu eggs and larvae by tetracycline-tagging of otoliths. Bulletin of the Japanese Society of Scientific Fisheries 51: 903–911.

Weber, D. D., and G. J. Ridgway. 1962. The deposition of tetracycline drugs in bones and scales of fish and its possible use for marking. Progressive Fish-Culturist 24:150–155.

Weber, D., and G. J. Ridgway. 1967. Marking Pacific salmon with tetracycline antibiotics. Journal of the Fisheries Research Board of Canada 24:849–865.

Wydoski, R., and L. Emery 1983. Tagging and marking. Pages 215–237 *in* L. A. Nielsen and D. L. Johnson, editors. Fisheries techniques. American Fisheries Society, Bethesda, Maryland.

American Fisheries Society Symposium 5:96–103, 1988

Abundance, Distribution, Growth, and Mortality of Larval Lake Herring in Western Lake Superior[1]

JAY T. HATCH

General College, University of Minnesota, Minneapolis, Minnesota 55455, USA

JAMES C. UNDERHILL

Department of Ecology and Behavioral Biology, University of Minnesota Minneapolis, Minnesota 55455, USA

Abstract.—Abundance, distribution, growth and mortality of larval lake herring *Coregonus artedii* were studied in the Duluth (Minnesota) to Superior (Wisconsin) area of western Lake Superior. Larvae were present in the area from late April to mid-June but were most abundant from mid- to late May. Normally, larvae were much more abundant in the upper 2 m of water than in the 4–6 m stratum. Mean densities of larvae in the vicinity of the Duluth harbor entry were 10–100 times greater than those in other areas and often exceeded 1,000/1,000 m^3 of water. Larvae remained in nearshore areas for 14–20 d after hatching; during this time they grew from about 10 mm standard length to 20 mm and reached the late mesolarval phase of development. Preliminary estimates of instantaneous growth rates for protolarval through mesolarval phases were 0.054–0.055, and total instantaneous mortality rates were 0.41–0.59. The Duluth–Superior area is an important developmental area for larval lake herring and offers excellent potential for developing larval abundance index stations.

Lake herring *Coregonus artedii* populations in Lake Superior have declined precipitously since the mid-1940s (Baldwin et al. 1979). Selgeby (1982) presented a convincing argument that, in the case of western Lake Superior populations, the decline occurred as a direct result of overexploitation. However, despite a great reduction in fishing intensity for these populations from the late 1960s to the present, lake herring abundance has remained relatively low. The reasons for the failure of western populations to recover are not clear. Anderson and Smith (1971) concluded that competition for zooplankton between the larvae of lake herring and rainbow smelt *Osmerus mordax* and, to a lesser extent, bloater *Coregonus hoyi* had a strong negative influence on lake herring abundance. More recently, Swenson and Heist (1981) suggested that rainbow smelt predation on larval lake herring could account for the lake herring's continued low abundance in the Duluth–Superior area. Selgeby et al. (1978) found that rainbow smelt predation was imporant in Black Bay, Ontario, but not in the Apostle Islands (Wisconsin) region. These studies suggest the hypothesis that high mortality during the larval period may be one reason that recruitment of adult lake herring has remained low. At the very least, these studies and others (Faber 1970; Lindström 1970; Reckahn 1970; Viljanen 1980) clearly point to the need for a greater understanding of the ecology of larval coregonines and of how their growth and mortality affects adult population size.

The present study was undertaken as a first step in elucidating the dynamics of larval lake herring populations in the Duluth–Superior area of western Lake Superior. Our immediate objective was to investigate the spatiotemporal distributions of lake herring larvae in the area so that index stations for measuring larval density, growth rates, and mortality rates could be established. We did not intend to estimate growth and mortality rates from this pilot study; however, since the project was not continued, we made preliminary estimates to determine if catch-curve analysis could be used for this purpose. We present data on the seasonal occurrence and distribution of lake herring in the study area, along with estimates of larval density in six index zones and the first field estimates of instantaneous growth and total mortality rates for larval lake herring.

Methods

Field sampling.—Sampling was conducted during May and June 1981 and April–June 1982 at various stations in the Duluth–Superior area of western Lake Superior (Figure 1). Sampling in 1981 was primarily exploratory; therefore, the

[1]Research Contribution 203 of Minnesota Sea Grant, University of Minnesota, St. Paul.

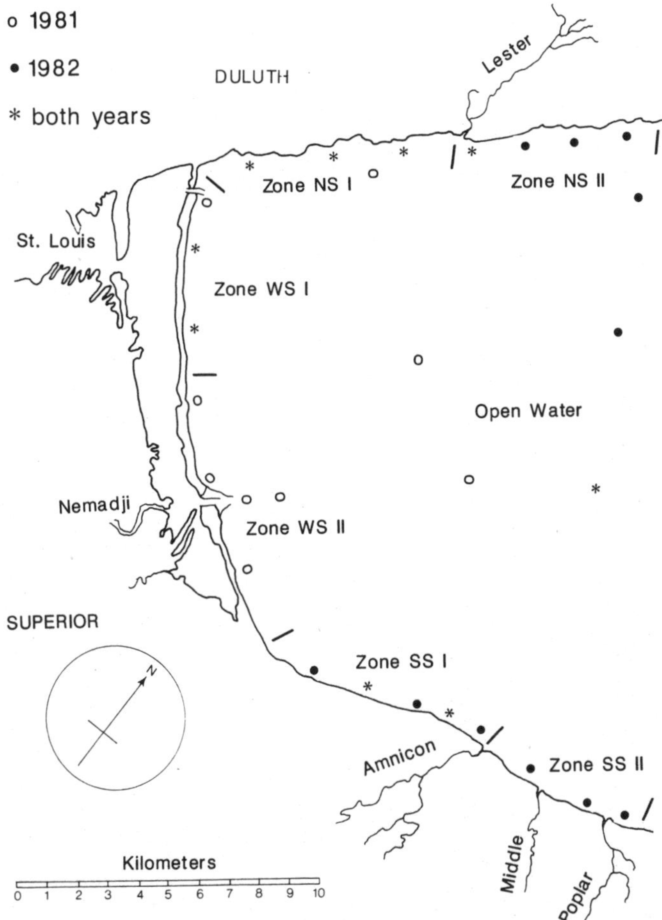

FIGURE 1.—Duluth–Superior area of western Lake Superior showing sampling stations and index zones where lake herring larvae were collected in 1981 and 1982. NS = north shore; WS = west shore, SS = south shore. The offshore station near zone NS II was considered an open water station.

spatial sampling on each excursion was not consistent. Based on the 1981 results, the study area was divided into six index zones, five of which were sampled systematically during each excursion in 1982 (zone WS II was not sampled in 1982). Two to four sets of samples usually were taken in a given sampling zone. Conclusions about the distribution of larval lake herring were drawn primarily from the 1982 data.

In 1981, larval lake herring were captured with a 0.5-m-diameter Bongo sampler fitted with one 200-μm-mesh and one 506-μm-mesh net (Smith and Richardson 1977) and a 1-m Tucker trawl of 506-μm-mesh netting (Tucker 1951). Paired sampling was conducted to determine the relative capture efficiencies of the two gear types and two mesh sizes. Paired *t*-tests showed no significant differences in catch per 1,000 m^3 between gear or between mesh sizes ($P > 0.10$). The Tucker trawl captured more total larvae per unit time than did the Bongo sampler, especially in areas of low abundance, and it allowed discrete depth sampling; therefore, we sampled exclusively with Tucker trawls in 1982.

Usually, two samplers were deployed simultaneously, one sampling the 0–2-m depth stratum and one the 4–6-m stratum. Depth of a sampler was determined from the angle and length of the towing cable. Sampling location and speed of the tow were determined by computer-assisted Loran C instrumentation. Water depth was determined hydroacoustically. The volume of water strained for each tow was determined with a calibrated General Oceanics model 2031 flowmeter. Measured tow speeds varied from 1.5 to 2.2 m/s. Volumes of water sampled averaged about 200

m^3/net for the Bongo sampler and 1,100 m^3 for the Tucker trawl. Water depths at the stations sampled in each of the index zones were NS I, 6–16 m; NS II, 6–32 m; WS I, 6–8 m; WS II, 6–15 m; SS I and SS II, 6–8 m. All specimens were preserved in Davidson's solution shortly after capture (Lam and Roff 1977).

Laboratory analysis.—All lake herring larvae were counted in most samples, and those in good condition were assigned to 1-mm standard length categories after being measured with a calibrated ocular micrometer. Larvae also were categorized as either protolarvae (larvae with no distinct median rays) or mesolarvae (larvae with at least one median fin ray but lacking a full complement of these rays) (Snyder 1976). Large samples (>500 larvae) were counted but then split using a Folsom zooplankton splitter to determine the distribution of standard length categories. Specimens that could not be assigned to a length category because they were in poor condition were counted and apportioned to length categories according to the percentage composition of the measured specimens in the sample. Catches were converted into numbers per 1,000 m^3 of water sampled.

Preliminary larval growth rates were estimated from the 1981 and 1982 catch data by the methods of Hackney and Webb (1978) and Cada and Hergenrader (1980). The catch of each length category was plotted against date of capture to determine the date on which the typical larva in each category attained its length. Then a plot of length against the derived dates yielded a growth curve that reflected the average growth rates of larvae during the sampling period. The curve is modeled by $L_t = L_0e^{Gt}$, where L_t is the length of the larva at end of the growth period, L_0 is the initial length, G is the coefficient of growth, and t is the length of the growth period in days. From this equation, the age of the midpoint of each length group can be determined and plotted against the $\log_e$-transformed catch data to yield a classical catch curve from which instantaneous mortality can be determined (Ricker 1975).

To gain some idea of age at a given stage of development and to have specimens of known parentage for identification purposes, fertilized eggs were obtained from Brule Lake, Minnesota, and Lake Superior (Apostle Islands) stocks and were incubated at the French River Hatchery (Minnesota Department of Natural Resources). All embryos were incubated through the winter in battery jars containing water from the French River. Incubation temperatures varied from 0.2 to 1.3°C. From early April through the end of the rearing sequence, Lake Superior water was used to maintain embryos and larvae. Larvae were fed brine shrimp continuously; periodically, zooplankton concentrated from Lake Superior water was added.

Results

Seasonal Abundance and Distribution

During 1981 and 1982, 8,013 and 8,354 lake herring larvae, respectively, were collected from western Lake Superior. Fifty-nine percent of these specimens were protolarvae (8.8–14.3 mm) and 41% were mesolarvae (12.6–22.4 mm). Metalarvae were not collected. Larvae were present in the study area on every sampling date, but their relative abundance was highly variable through space and time (Table 1). In 1981, mean catch remained near or above 200/1,000 m^3 throughout May but fell sharply to 18/1,000 m^3 by 4 June. In contrast, mean catches in 1982 remained well below 100/1,000 m^3 until 24 May and peaked around 28 May. The mean catch remained above 100/1,000 m^3 through 3 June but fell sharply to 4/1,000 m^3 by 11 June.

In 1982, the first major hatching series of larval lake herring occurred sometime between 23 April and 7 May. Only two yolk-sac protolarvae (9.7–10.3 mm) were captured on 23 April, but by 7 May advanced mesolarvae (17–22 mm) appeared in the catches along with new yolk-sac protolarvae indicating the presence of two or more hatching cohorts at this time (Figure 2). We did not sample prior to 7 May in 1981, but the length-frequency distribution on that date suggests that the hatching series was several days ahead of the 1982 sequence (Figure 3). In 1981, there was virtually no further recruitment of yolk-sac protolarvae after 7 May. However, in 1982, substantial recruitment of these young larvae continued to the end of May. Larval lake herring were present in the study area on the last sampling date of each year, but they were few in number. The movement out of the study area in June was not size specific, suggesting that some physical or biotic factor unrelated to ontogeny caused the final emigration.

Because of the exploratory nature of sampling in 1981, statistical comparisons of distribution patterns were not carried out. However, sampling results suggested three trends in the distribution of lake herring larvae. First, lake herring larvae were not present in deep open water. Second, larvae were most abundant in the area around the

TABLE 1.—Average number of larval lake herring caught per 1,000 m^3 of water from seven zones in western Lake Superior, 1981 and 1982. Stations sampled within the zones were not the same in both years (see Figure 1). NS = north shore; WS = west shore; SS = south shore; S = 0–2-m depth strata; D = 4–6-m depth strata. Unless otherwise indicated, all samples were taken during daylight hours.

Sampling dates	Number of tows	Number of larvae by sampling zone and depth													
		NS I		NS II		WS I		WS II		SS I		SS II		Open water	
		S	D	S	D	S	D	S	D	S	D	S	D	S	D
1981															
7 May	15							382						0	0
15 May	29	706		0		1,201		755	198					0	0
21 May	18							2	6					0	0
28 May	26	1,111	23			102	104	31	28						
28 May[a]	8	312				87									
4 Jun	17	12	2			16	0	17	5	30	13				
1982															
23 Apr	3			<1	0	<1	0			0	0				
7 May	20	16	11	1	6	7	1			15	0	0	0		
14 May	17	59	56	0	23	13	16			17	0			0	0
24 May	24	493	119	3	154	59	0			7	<1	14	3		
28 May	13	1,118	49			24	4			0	0	0	0	0	0
30 May	14			363	196	2	0			4	0	2	0		
3 Jun	24	596	92	10	34	51	8			1	0	1	0		
11 Jun	24	0	0	1	7	0	0			<1	0	2	0		
18 Jun	20	1	0	4	3					0	0	<1	0	4	<1

[a]Nighttime samples.

Duluth harbor entry, which included parts of zones NS I and WS I. Third, the depth distribution of larvae was not the same in all zones.

These trends largely were borne out by the systematic sampling in 1982. Again, lake herring were not found at open, deepwater stations, except on 18 June when 14 larvae were collected from three stations. Larvae consistently were most abundant along the north shore between the Duluth harbor entry and the Lester River (zone NS I), where individual catches exceeded 1,000/1,000 m^3 on six occasions. Lake herring became less abundant further up the north shore (zone NS II) and to the west of the Duluth entry (zone WS I), and their abundance was consistently lowest along the south shore (zones SS I and SS II). A comparison of the north shore and south shore catch frequencies, based on all surface and depth samples on all dates, illustrated more convincingly the disparity in abundance between the two areas (Figure 4). Contingency table analysis of the zero and nonzero catch frequencies along the two shores was highly significant ($\chi^2 = 23.41$; df = 1; $P < 0.001$). A similar comparison of zones NS II and WS I was not significant ($\chi^2 = 1.55$; df = 1; $P > 0.20$). Finally, larval lake herring clearly were more abundant in the 0–2-m-depth stratum than in the 4–6-m stratum, except in zone NS II, where mean catches suggested that the reverse was true. A contingency table analysis of catches at two depths in the two north shore zones demonstrated the statistical significance of this pattern ($\chi^2 = 1128$; df = 1; $P < 0.001$).

Growth and Mortality

From the data in Table 2, which compares developmental stages of our wild-caught and reared larvae and associates them with age, and from comparisons with the studies of Pritchard (1930) and Hinrichs and Booke (1975), we estimated that on average lake herring larvae spent about 14–20 d in the study area after hatching. During this time, they reached the late mesolarval phase of development and attained sizes of 17–22 mm standard length.

We attempted preliminary estimates of instantaneous growth and mortality rates using the 1981 total catch data. After plotting the catch for each 1-mm length-group on semilog paper, it was apparent that our data would not yield reasonable results for length groups below 11 mm or above 16 mm. From our length-frequency data and our reared specimens, we estimated that lake herring larvae hatched at about 10 mm and took 1–2 d to reach swim-up (John and Hasler 1956; Hinrichs and Booke 1975). Thus, it makes sense that the 11-mm length category would be the first fully recruitable size-group. The catch of specimens over 16 mm was simply too sporadic to yield reasonable plots for these large sizes. However,

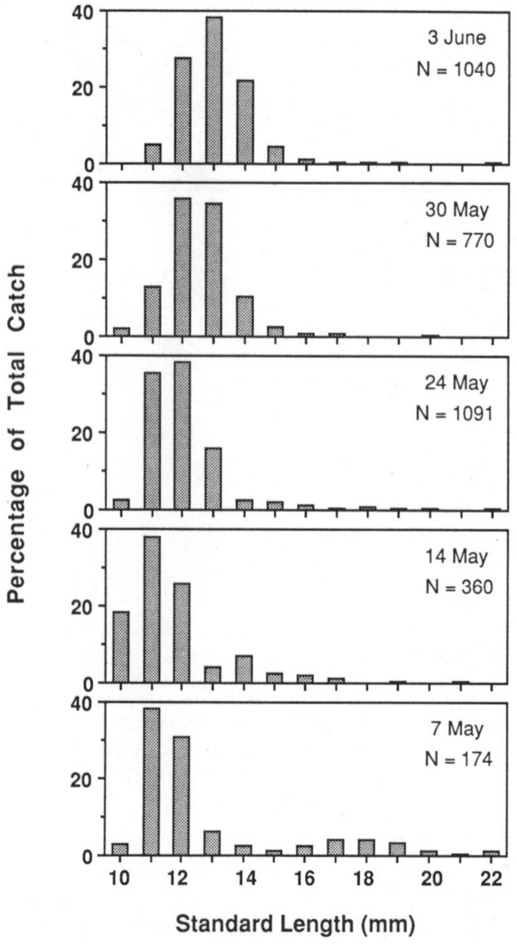

FIGURE 2.—Length-frequency distributions of larval lake herring collected in the Duluth–Superior area of Lake Superior in 1982; all samples combined. Distributions are not shown for 11 and 18 June because fewer than 30 larvae were caught on each of those dates.

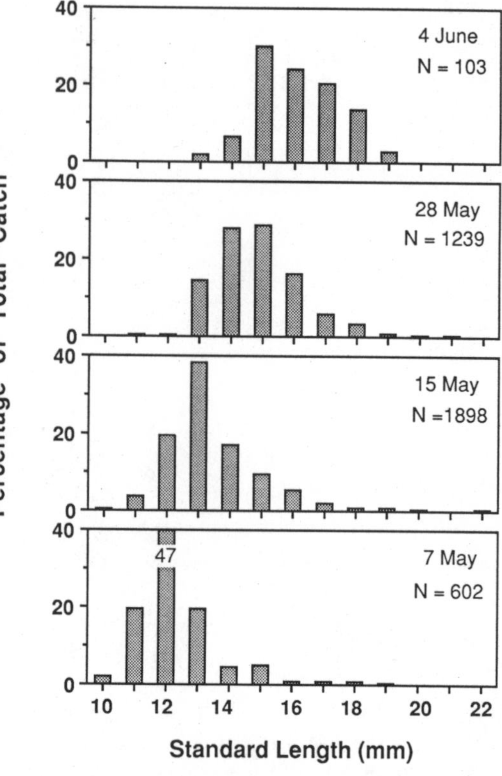

FIGURE 3.—Length-frequency distributions of larval lake herring collected in the Duluth–Superior area of Lake Superior in 1981; all samples combined.

for larvae in the 11–16-mm length categories we computed the following exponential growth curve: $L = 9.90e^{0.0545t}$; L is standard length in millimeters and t is age in days ($r = 0.994$).

From this equation, we determined the average age of larvae as they attained the midpoint length within each length-group between 11 and 16 mm. Although we used whole-number 1-mm length-groups in our graphs, the actual midpoint lengths were used in all calculations. The midpoints were 10.60, 11.75, 12.70, 13.80, 14.85, and 15.90. After the age in days had been assigned to each of these midpoints, we plotted $\log_e$ of cumulative catch (the area under a catch versus sampling date plot) against age to give a larval catch curve. The slope of this curve was -0.4110 ($r = 0.994$), which means that the instantaneous total mortality rate, Z, has the value of 0.4110 when age is expressed in days. A rather similar estimate of Z was obtained when a grand average of catch was used instead of cumulative catch. The Z-value of this catch curve was 0.4176 ($r = 0.992$).

We attempted one more set of calculations based on the 1982 catch data from zone NS I, the zone of highest density throughout May and early June. From these data, we computed the growth curve, $L = 10.17e^{0.0543t}$; $r = 0.974$. A catch curve based on cumulative catch yielded $Z = 0.5837$ ($r = 0.970$), while one based on grand average catch yielded $Z = 0.5931$ ($r = 0.958$).

Discussion

The seasonal abundance pattern of protolarval and mesolarval lake herring in western Lake Superior appears to be a fairly stable one that varies only slightly geographically or from year to year. Larvae are present primarily during May and the first half of June with peak abundance

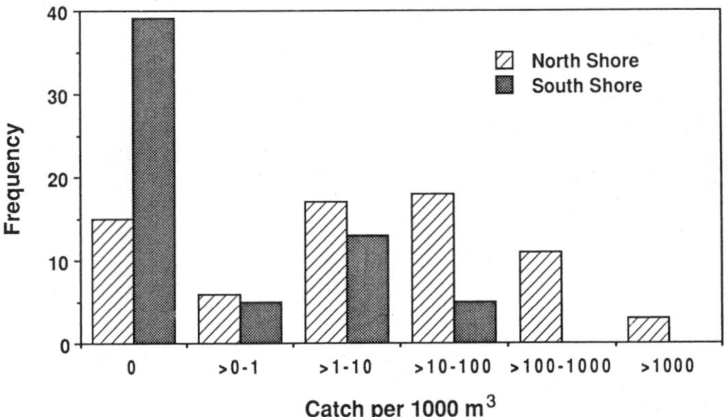

FIGURE 4.—Comparison of lake herring larval abundance in the north shore and south shore index zones of western Lake Superior in 1982 based on frequencies of catch categories (number of larvae/1,000 m^3 of water). The use of catch frequencies provides another way to view highly variable and skewed field data.

occurring in mid to late May. The patterns we observed were very similar to the ones reported by Anderson and Smith (1971) for larvae collected in the Duluth–Superior and the Apostle Islands areas in 1967 and 1968 and to the ones reported by Swenson and Heist (1981) for the Duluth–Superior area in 1979 and 1980. Selgeby et al. (1978) reported similar seasonal patterns for larvae collected in the Apostle Islands area and, to a lesser extent, in Black Bay in 1974. Such stability is to be expected since the thermal regimes experienced during incubation probably fluctuate very little geographically or temporally. In testing a series of incubation thermal regimes, which covered all of the possibilities reasonable for Lake Superior, John and Hasler (1956) produced no more than a 7-d difference in hatching dates. Thus, future attempts to monitor larval abundance and measure growth and mortality could be framed adequately around a weekly sampling schedule beginning in late April and ending in mid-June of each year.

TABLE 2.—Size comparisons of wild-caught larval lake herring during five phases of development with reared larvae of known age. Reared larvae were from Brule Lake and Lake Superior stocks. Wild-caught larvae were collected from western Lake Superior, 1981 and 1982.

Developmental phase	Standard length range (mm)		Age range of reared larvae (d)
	Wild-caught larvae	Reared larvae	
Early protolarvae	8.8–12.2	9.5–11.9	0–6
Late protolarvae	11.5–14.3	10.3–12.4	7–9
Early mesolarvae	12.6–17.2	10.8–14.0	10–21
Middle mesolarvae	14.0–21.0	13.3–17.5	21–39
Late mesolarvae	19.1–22.4	15.3–18.3	43–61

Making strict comparisons of actual density estimates among studies is probably unrealistic. Density estimates derived from different sampling techniques usually cannot be reconciled based on the data reported, and in most cases, the high variability of catches within a study makes estimates of mean density very difficult to analyze statistically (Hamley et al. 1983). However, general comparisons of orders of magnitude may provide a rough way to compare larval density estimates. The highest estimates from this study are of the same order of magnitude (1,000/1,000 m^3) as those from Black Bay and as much as two orders of magnitude above those from the Apostle Islands (Selgeby et al. 1978) and the high-density area of Oneida Lake (Clady 1976). Based on these comparisons, we believe that the north shore area in the vicinity of the Duluth harbor entry is an important location for the early development of lake herring in western Lake Superior, especially when we consider that our density estimates probably are low because we were unable to sample water less than 5 m deep. Pritchard (1930), Faber (1970), and Clady (1976) all reported the highest relative abundance of lake herring larvae near shore in water less than 2 m deep.

The growth rate estimates from the 1981 and 1982 data agree rather closely with each other, but we believe that they are overestimates. If we use the 1982 data to calculate average daily growth increment of the catch during the first week after hatching, we get a mean of 0.69 mm/d. Anderson and Smith (1971) reported increments of 0.13–

0.63 mm/d (converted from average weekly increments for lake herring larvae reared from Duluth–Superior area stock). Their larvae grew to 37.5 mm total length at simulated lake temperatures (6.1–10.5°C). The grand average increment was 0.27 mm/d. Our specimens experienced slightly lower temperatures in Lake Superior, and thus it seems unlikely that they would have grown 0.69 mm/d, especially in their first week. Hinrichs (1979) did not report weekly or daily increments for the larvae he reared, but calculations from his data yield consistent values of 0.25 mm/d during the first 16 d. These larvae were reared at temperatures of 9.9–13.0°C, which are close to the optimum growth temperatures reported for this species by McCormick et al. (1971).

It is still possible that the field specimens could have grown faster than reared specimens had the quality and quantity of food been poorer for reared specimens. Exogenous nutrition is extremely important for good growth even during the yolk-sac phase (John and Hasler 1956). Taylor and Freeberg (1984) found that larval lake whitefish *Coregonus clupeaformis* fed at densities of 1.8 zooplankton/fish grew 28% slower than those fed at 110 zooplankton/fish. The specimens that we reared fed poorly, and their average daily increments in standard length during the first week were 0.08–0.13 mm, with a grand average of 0.11 mm/d. The highest incremental growth rate determined for an individual was 0.23 mm/d for a 32-d-old larva (using 10.0 mm standard length as a mean size at hatching). When length ranges of our field and hatchery specimens are compared at similar stages of development, it is clear how much growth rates can be affected by food availability (Table 2).

If we compare our incremental rates from the field with those of other field estimates, ours still seem high. Clady (1976) reported a value of 0.6 mm/d for Oneida Lake larvae, but this was an average over a size range of 11–21 mm total length and also included data from 6 years. Oneida Lake also is warmer and much more productive than Lake Superior. Growth conditions in the Bay of Quinte, Lake Ontario, are also more favorable than those of Lake Superior, and lake herring larvae there showed an average daily increment of 0.58 mm during a 7-d period in early May. We calculated this value from mean length of specimens captured each day as reported by Pritchard (1930). However, if we use Pritchard's estimate that 16-mm larvae were about 20 d old, the daily increment would be about 0.3 mm (using his estimate of 10 mm as an average hatching size).

It is clear from the size-specific catch data in both years that we undersampled newly hatched larvae. Such underrepresentation in the catch would contribute to an underestimation of age and, thus, an overestimation of growth rate. The same bias would lead to an underestimation of mortality rate. However, a loss of larger larvae from the catch, due either to avoidance or emigration, could have overcompensated for this bias and given us an unrealistically high mortality rate. There are no other field estimates of larval lake herring mortality in the literature, but rates for other freshwater species generally have been below 0.4 (Hachney and Webb 1978; Cada and Hergenrader 1980). Essig and Cole (1986) calculated values for larval alewives *Alosa pseudoharengus* of 0.097–0.596 using the same database but different methods of calculation, which suggests that values derived from different methods are not directly comparable.

Although this was a preliminary study, we feel justified in concluding that the Duluth–Superior area of western Lake Superior offers excellent potential for developing index stations to measure larval lake herring relative abundance, growth, and mortality. If larval abundance is a reliable predictor of subsequent year-class strength, fishery managers will be able to estimate the status of fishing stocks 3–4 years sooner than is now possible. Further, if systematic estimates of growth and mortality can be made year to year, biologists could better assess which extrinsic factors lead to good relative growth and low mortality in larval populations. Moreover, lake herring larvae can be sampled with relatively little effort because they are abundant and remain near shore in a relatively small portion of the Duluth–Superior area for at least 20 d.

Acknowledgments

We thank Mary Balcer and the captain, David Anderson, and crew of the Research Vessel *L. L. Smith, Jr.*, for their expert assistance in the field, and Herb Johnson and the staff of the French River Hatchery for their help in rearing lake herring larvae. We acknowledge the many significant contributions made throughout this study by Theodore Halpern. William D. Schmid provided advice on statistics. This research was supported by Minnesota Sea Grant Program, grant NA81AA/0/00114, project R/F-8.

References

Anderson, E. D., and L. L. Smith, Jr. 1971. Factors affecting abundance of lake herring (*Coregonus artedii* Lesueur) in western Lake Superior. Transactions of the American Fisheries Society 100:691–707.

Baldwin, N. S., R. W. Saalfeld, M. A. Ross, and H. J. Buettner. 1979. Commercial fish production in the Great Lakes, 1867–1977. Great Lakes Fishery Commission Technical Report 3.

Cada, G. F., and G. L. Hergenrader. 1980. Natural mortality rates of freshwater drum larvae in the Missouri River. Transactions of the American Fisheries Society 109:470–483.

Clady, M. D. 1976. Distribution and abundance of larval ciscoes, *Coregonus artedii*, and burbot, *Lota lota*, in Oneida Lake. Journal of Great Lakes Research 2: 234–247.

Essig, R. J., and C. F. Cole. 1986. Methods of estimating larval fish mortality from daily increments in otoliths. Transactions of the American Fisheries Society 115:28–33.

Faber, D. J. 1970. Ecological observations on newly hatched lake whitefish in South Bay, Lake Huron. Pages 481–500 *in* C. C. Lindsey and C. S. Woods, editors. Biology of coregonid fishes. University of Manitoba Press, Winnipeg, Canada.

Hackney, P. A., and J. C. Webb. 1978. A method for determining growth and mortality rates of ichthyoplankton. Pages 115–124 *in* L. D. Jensen, editor. Fourth national workshop on entrainment and impingement. Ecological Analysts Communications, Melville, New York.

Hamley, J. M., T. P. Howley, and A. L. Punhani. 1983. Estimating larval fish abundances from plankton net catches in Long Point Bay, Lake Erie, in 1971–78. Journal of Great Lakes Research 9:452–467.

Hinrichs, M. A. 1979. A description and key to the eggs and larvae of five species of fish in the subfamily Coregoninae. Master's thesis. University of Wisconsin, Stevens Point.

Hinrichs, M. A., and H. E. Booke. 1975. Egg development and larval feeding of the lake herring, *Coregonus artedii* (LeSueur). Museum of Natural History, University of Wisconsin, Reports on Fauna and Flora of Wisconsin 10(4):75–86.

John K. R., and A. D. Hasler. 1956. Observations on some factors affecting the hatching of eggs and the survival of young shallow water cisco (*Leucichthys artedii* LeSueur) in Lake Mendota, Wisconsin. Limnology and Oceanography 1:176–194.

Lam, C. N. H., and J. C. Roff. 1977. A method for separating alewife *Alosa pseudoharengus* from gizzard shad *Dorosoma cepedianum* larvae. Journal of Great Lakes Research 2:313–316.

Lindström, T. 1970. Habitats of whitefish in some north Swedish lakes at different stages of life history. Pages 461–479 *in* C. C. Lindsey and C. S. Woods, editors. Biology of coregonid fishes. University of Manitoba Press, Winnipeg, Canada.

McCormick, J. H., B. R. Jones, and R. F. Syrett. 1971. Temperature requirements for growth and survival of larval ciscos (*Coregonus artedii*). Journal of the Fisheries Research Board of Canada 28:924–927.

Pritchard, A. L. 1930. Spawning habits and fry on the cisco (*Leucichthys artedi*) in Lake Ontario. Contributions to Canadian Biology and Fisheries 6:227–240.

Reckahn, J. A. 1970. Ecology of young lake whitefish (*Coregonus clupeaformis*) in South Bay, Manitoulin Island, Lake Huron. Pages 437–460 *in* C. C. Lindsey and C. S. Woods, editors. Biology of coregonid fishes. University of Manitoba Press, Winnipeg, Canada.

Ricker, W. E. 1975. Computation and interpretation of biological statistics of fish populations. Fisheries Research Board of Canada Bulletin 191.

Selgeby, J. H. 1982. Decline of lake herring (*Coregonus artedii*) in Lake Superior: an analysis of the Wisconsin herring fishery, 1936–78. Canadian Journal of Fisheries and Aquatic Sciences 39:554–563.

Selgeby, J. H., W. R. MacCallum, and D. V. Swedberg. 1978. Predation by rainbow smelt (*Osmerus mordax*) on lake herring (*Coregonus artedii*) in western Lake Superior. Journal of the Fisheries Research Board of Canada 35:1457–1463.

Smith, P. E., and S. L. Richardson. 1977. Standard techniques for pelagic fish eggs and larva surveys. FAO (Food and Agriculture Organization of the United Nations) Fisheries Technical Paper 175.

Snyder, D. E. 1976. Terminologies for intervals of larval fish development. United States Fish and Wildlife Service Biological Services Program, FWS/OBS–76/23:41–58.

Swenson, W. A., and B. Heist. 1981. Optimizing yield from western Lake Superior commercial fisheries through smelt stock assessment. University of Wisconsin, Sea Grant Institute, Final Report R/LR-8, Madison.

Taylor, W. W., and M. H. Freeberg. 1984. Effect of food abundance on larval lake whitefish, *Coregonus clupeaformis* Mitchill, growth and survival. Journal of Fish Biology 25:733–741.

Tucker, G. H. 1951. Relation of fishes and other organisms to the scattering of underwater sound. Journal of Marine Research 10:215–238.

Viljanen, M. 1980. A comparison of a large diameter corer and a new hydraulic suction sampler in sampling eggs of *Coregonus albula*. Annales Zoologici Fennici 17:269–273.

American Fisheries Society Symposium 5:104–109, 1988

Movements of Young Lake Sturgeons Stocked in the Menominee River, Wisconsin

THOMAS F. THUEMLER

*Wisconsin Department of Natural Resources, Industrial Parkway, Box 16
Marinette, Wisconsin 54143, USA*

Abstract.—The Wisconsin Department of Natural Resources is attempting to reestablish a population of lake sturgeon *Acipenser fulvescens* in a section of the Menominee River. Yearling hatchery-reared sturgeons of Lake Winnebago stock and native transplanted Menominee River fish were released into this section of the river. Twenty-four of these fish were tagged with radio transmitters to monitor their movement patterns. Fish of Lake Winnebago origin moved rapidly downstream, as far as 32 km in 24 h. Most of these fish moved out of the intended section of river and into downstream areas. Lake sturgeons transplanted from another section of the Menominee did not exhibit the same rapid downstream movement. These fish stayed in the intended stretch of river, and most did not move downstream. Results indicated that juvenile lake sturgeons from a riverine source should be used to repopulate the Menominee River.

The lake sturgeon *Acipenser fulvescens* is listed as threatened in the USA (Miller 1972) and is on the watch list in Wisconsin (Les 1979). The Menominee River, which forms the border between northeastern Wisconsin and the Upper Peninsula of Michigan, contains one of the few remaining fishable stocks of lake sturgeon in either state (Thuemler 1985). A detailed study of lake sturgeon populations in the Menominee River was done in the early 1970s (Priegel 1973). Although a 39-km stretch of the upper Menominee, known as Sturgeon Falls, historically contained lake sturgeons, they have since disappeared from there probably owing in part to pollution. Water quality has dramatically improved since the early 1970s, and along with this improvement came a rejuvenation of the sport fishery for species other than lake sturgeon. Species such as walleye *Stizostedion vitreum*, smallmouth bass *Micropterus dolomieui*, and northern pike *Esox lucius* have repopulated this section of the river. Lake sturgeons are still not present in the area because downstream dams form a barrier to upstream movement. Lake sturgeons were never found in the river above Sturgeon Falls; therefore, downstream drift is impossible.

Wisconsin scientists have been raising lake sturgeons experimentally since 1979 (Folz et al. 1983). The Sturgeon Falls section of the Menominee was considered to be a good location to check on the survival of these hatchery-reared fish and to attempt to reestablish the lake sturgeon population. Lake sturgeons have been stocked in this section of the river since 1982 (Thuemler 1985). However, electrofishing surveys conducted in 1984 and 1985 recovered only one lake sturgeon in the stocked section of river. The stocked fish are believed to have moved out of the intended section of river and into downstream areas.

Radio telemetry was used in this study to check the movement patterns of young lake sturgeons immediately after their release into the Sturgeon Falls section of the Menominee River.

Study Area

The Menominee River is formed by the Brule and Michigamme rivers and flows in a southeasterly direction for 152 km before entering the waters of Green Bay (Figure 1). The river water is hard, slightly alkaline, and stained (Carlson et al. 1975); the river averages 146 m wide and one m deep in the study area. Fish inhabiting the river, in addition to lake sturgeon, include walleye, smallmouth bass, northern pike, largemouth bass *Micropterus salmoides*, channel catfish *Ictalurus punctatus*, rock bass *Ambloplites rupestris*, yellow perch *Perca flavescens*, white sucker *Catostomus commersoni*, common carp *Cyprinus carpio*, some redhorse *Moxostoma* spp., and a large variety of coarse fish (Thuemler 1985).

The study area extends 119 km from the Sturgeon Falls dam downstream to the Upper Scott flowage (Figure 1). Five hydroelectric dams are located within this reach of river: Sturgeon Falls at the upper end, Chalk Hills at 39 km downstream, White Rapids at 42 km, Grand Rapids at 85 km, and Upper Scott at 119 km.

Two natural falls are also located in the study area, Quiver Falls and Pemene Falls, 16 and 24 km downstream from Sturgeon Falls, respectively. These falls are not a barrier to movement

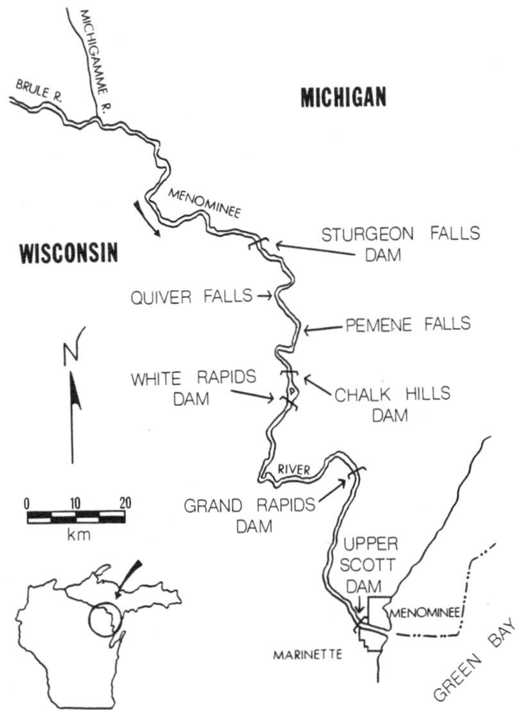

FIGURE 1.—Lake sturgeon study area on the Menominee River, Wisconsin—Michigan boundary water.

of adult lake sturgeons; however, they are a barrier to upstream movement of the subadult fish used in this study.

Average annual discharge of the Menominee River in this section is 83.8 m^3/s (Holmstrom 1980). Mean monthly flow for the October study period was 71.8 m^3/s (Holmstrom, U.S. Geological Survey, personal communication).

Methods

Lake sturgeons used in the study in 1985 were raised to 18 months of age at a state hatchery in Wild Rose, Wisconsin. Eggs were taken from adult lake sturgeons of Lake Winnebago stock in April 1984 on the Fox River, Wisconsin (Czeskleba et al. 1985). Lake sturgeons used in the study in 1986 were obtained by electrofishing the Menominee River immediately below the Grand Rapids dam on September 29, 1986.

A test was conducted in September 1985 to determine if lake sturgeons could tolerate the implant of a radio transmitter in their body cavity. Dummy radios, the size and weight of the actual transmitters, were implanted into the abdominal cavities of six lake sturgeon yearlings on September 3, 1985. These fish averaged 29.9 cm long and 121 g in weight. All surgical tools were sterilized with a 1:750 aqueous solution of benzalkonium chloride and then rinsed with a sterile saline solution. Fish were anesthetized with tricaine (MS-222). A 15–20-mm midventral incision was made in the abdominal wall anterior to the vent. Dummy transmitters were placed in the body cavities after being immersed in the benzalkonium chloride and saline solutions. The 15-cm antenna wire of each transmitter protruded out through a needle hole in the body wall posterior to the incision. Five or six sutures were taken in each incision with a surgical cutting needle and nonabsorbable nylon suture. The operations took between 7.5 and 13 min/fish. After surgery, 0.1 cm^3 of oxytetracycline (50 mg oxytetracycline/mL) was injected into the body cavity of each fish. The fish were then placed in a tank where they recovered from anesthesia within 15 min.

Condition of the six fish with the dummy transmitters was monitored until October 3, 1985. Incisions healed completely in 10 d to 2 weeks. No mortalities occurred over the month that the fish were held in the hatchery. Lake sturgeons appeared to be in good condition, and their behavior was not visibly affected by the implants.

Radio transmitters used in this study were designed and assembled by Advanced Telemetry Systems. They were all single-stage units encapsulated in clear waterproof resin with flexible 15-cm-long teflon whip antennas. They were on the 50 MHz band spaced 0.01 MHz apart. The transmitters weighed approximately 3.5 g, measured, 17 × 9 × 6 mm, and had an expected life of 20 d.

The radio transmitters were implanted in 14 yearling lake sturgeons of Lake Winnebago origin on October 3, 1985, by the same procedures used with the dummy implants. The 14 fish averaged 30.5 cm in length (range, 28–32 cm) and 140 g in weight (range, 96–167 g). They were held in hatchery raceways until they were released in the Menominee River on October 7, 1985.

The 10 Menominee River lake sturgeons used in the study in 1986 were implanted with transmitters on October 1, 1986. These fish had an average length of 48 cm (range, 40.1–51.3 cm) and a weight of 376 g (range, 226–455 g). They were held in a fish tank overnight prior to being released in the Menominee River on October 2, 1986.

Lake sturgeons in both years were stocked immediately below the Sturgeon Falls dam. Fish were tracked daily throughout the expected life of the transmitter. A hand-held loop antenna and a Yagi antenna were used to track the lake sturgeons. The study area was surveyed for tagged fish daily, and locations of lake sturgeons were noted by 0.4-km sections of the river.

Results

Lake Sturgeons Released in 1985

Tagged lake sturgeons were released at 1400 hours on October 7, 1985, near the Sturgeon Falls dam. Signals from four of these lake sturgeons were never received after their release, possibly because the implanted radios were defective. Although each transmitter was tested twice, once at implantation and again just prior to the fish's release, the reliability of these small single-stage units is less than desirable. Some reliability had to be sacrificed to obtain transmitters small enough to be used in these small fish.

The remaining 10 fish gradually moved downstream; 4 h after their release, all fish remained within 2 km of the stocking site. By the next day, all 10 fish were at least 11 km downstream of Sturgeon Falls, and 2 fish were located 38 km downstream in the Chalk Hills flowage. On October 9, 6 radio-tagged lake sturgeons were found in the vicinity of the Chalk Hills flowage. Over the next 12 days, 76 contacts were made with the 10 fish. Nine of these fish had moved downstream at least as far as the Chalk Hills flowage, whereas the remaining fish was contacted only once, 16 km downstream of the stocking site. Five radio-tagged lake sturgeons moved over the Chalk Hills dam and into the White Rapids flowage. One of these fish moved 119 km downstream from Sturgeon Falls, passing over three dams, to the Upper Scott flowage in just 8 d.

A typical movement pattern was exhibited by fish 1253 (Figure 2). This fish moved downstream to the upper end of the Chalk Hills flowage and then upstream to the pool immediately below Pemene Falls. Evidently the falls posed a barrier to further upstream movement of these small fish. Upon last contact, seven of the lake sturgeons were found in a flowage environment (Figure 3). Only one fish was located on October 21, 14 d after stocking, and that signal was very weak. No fish were found after that date. The fastest recorded downstream travel was 39 km/d, the fastest upstream was 3.2 km/d.

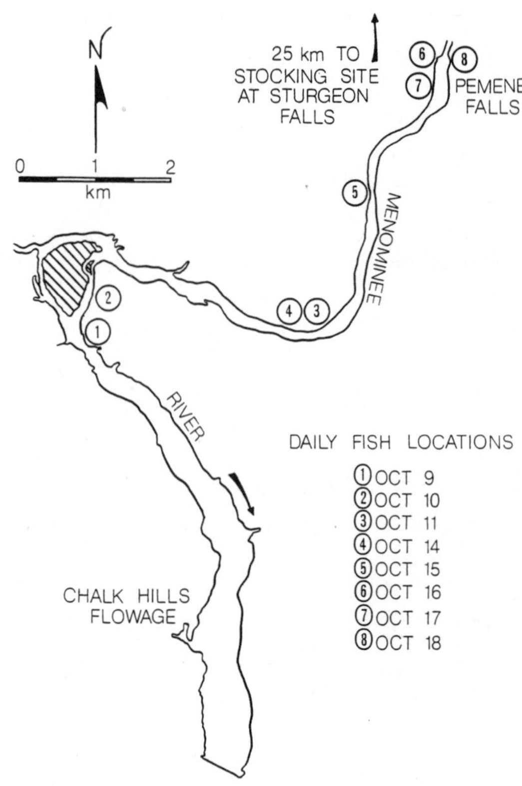

FIGURE 2.—Movement pattern of a radio–tagged lake sturgeon (fish 1253) stocked in the Menominee River on October 7, 1985.

Lake Sturgeons Released in 1986

In 1986, 10 lake sturgeons tagged with radio transmitters were released at 1040 hours on October 2 near the Sturgeon Falls dam. Four hours after their release, 9 of the tagged lake sturgeons remained within 2 km of the stocking site. One fish was never contacted after release, perhaps because it had a defective transmitter. Five lake sturgeons remained within 5 km of the release site on the next day. In general, fish dispersed much slower than in 1985 and did not travel as far. Three fish stayed within 3 km of the release site for 3 d, but another fish stayed within 1 km of Sturgeon Falls throughout the entire 19-d monitoring period. Overall, 80 contacts were made on nine fish over 19 d. Upon last contact, only one of the nine tagged fish was found in a flowage environment (Figure 3). The other eight fish stayed in the intended section of the river, six of them near the mouth of the PemeBonWon River. A typical movement pattern of the lake sturgeons stocked in 1986 was exhibited by fish 1335 (Figure 4). The fastest any lake sturgeon moved downstream in

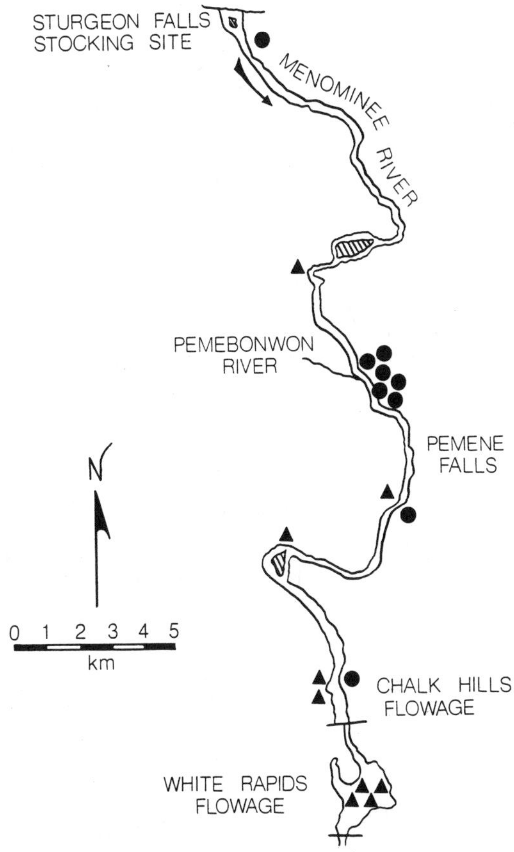

FIGURE 3.—Locations where individual radio-tagged lake sturgeon were last contacted in the Menominee River in 1985 (▲) and 1986 (●). In 1985, one fish was last located in Upper Scott flowage, 119 km downstream from the stocking site.

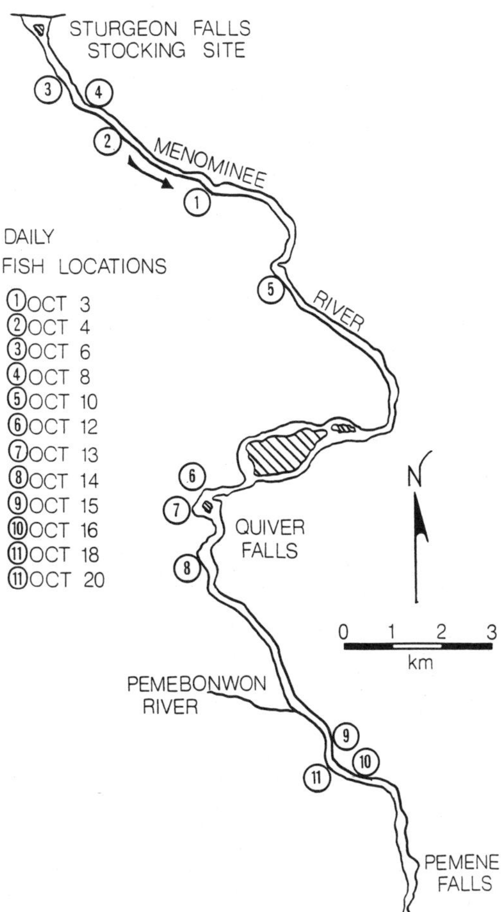

FIGURE 4.—Movement pattern of a radio-tagged lake sturgeon (fish 1335) stocked in the Menominee River on October 2, 1986.

1986 was 14.5 km/d, and the fastest upstream movement was 5 km/d. No lake sturgeons were contacted on October 23, 1986, and all tracking was discontinued from that date onward.

Discussion

The lake sturgeon is a typical inhabitant of large rivers and lakes throughout its range (Scott and Crossman 1973; Becker 1983). Seasonal movement patterns are not well known outside of spawning migrations (Scott and Crossman 1973). At spawning time, lake sturgeons leave their home areas, move upstream to spawn, and then return to their home areas (Scott and Crossman 1973; Priegel and Wirth 1974; Threader and Brousseau 1986). Strong homing tendencies have been described particularly for Lake Winnebago lake sturgeons (Priegel and Wirth 1975; Folz and Meyers 1985).

Lake sturgeons in the Menominee River spend their entire life in a riverine habitat (Priegal 1973; Thuemler 1985). They move to upstream spawning areas, generally within 30 km of their home area, and then return to their home areas after spawning (Thuemler 1985). Lake sturgeons tagged in this study in 1985 were of Lake Winnebago origin. When released, they moved rapidly downstream. Nine of the 10 fish moved at least 39 km downstream. Seven of these nine fish were eventually found in a flowage type environment. These fish could have been seeking the lacustrine type habitat of their parents. Transplanted Menominee River fish used in 1986 did not show the same rapid downstream movement upon release. Only one fish traveled as far as 39 km downstream from the stocking site, and eight of the nine fish remained in a riverine habitat.

Several possible explanations exist for the differences observed in the movement of released fish. River flows, water temperature, available food, genetic origin, and age of stocked fish all might influence fish movement. The study was conducted during the midfall in both years. River flows were monitored at a U.S. Geological Survey station located within the study area. Although flows fluctuated greatly during the study period, the fluctuations in both years were similar. Fish movement was not correlated with either increasing or decreasing flows. Water temperatures ranged between 9 and 16°C in both years. There was no correlation between water temperature and movement of the radio-tagged lake sturgeons. Little is known about the food habits of small lake sturgeons in the Menominee River. Presumably they feed primarily on crayfish, molluscs, and insect larvae. There was no shortage of this type of food in the study area and no reason to believe that there would be any difference in food abundance between the 2 years of the study. There was a difference in the age of the fish used in 1985 and 1986. The fish released in 1985 were all yearlings, about 18 months of age. In 1986, the lake sturgeons were not aged but, based on the size of the fish, they were either 2 or 3 years old. In surveys conducted on sections of the Menominee River that have self-sustaining lake sturgeon stocks, we see lake sturgeons up to 100 cm in length inhabiting the same type of river habitat. I do not believe that the different movement patterns exhibited by the lake sturgeons released in 1985 and 1986 were due to the differences in size or age of the lake sturgeons stocked.

I believe that the genetic differences within the Lake Winnebago and the Menominee River stocks of lake sturgeon are the probable causes for the movement differences seen in this study. Preliminary work on the genetic structure of different strains of white sturgeon *Acipenser transmontanus* has already been done (Bartley et al. 1985). Genetic variability, measured as average heterozygosity, in lake sturgeon is greater than that reported for other species of sturgeon (D. Bartley, University of California, Davis, personal communication). Migratory behavior is not known to be a genetic trait in lake sturgeon; however, Rounsefell (1958) believed that anadromy was a genetically linked trait in some strains of landlocked Atlantic salmon *Salmo salar*.

More data on the genetic structure of natural populations of lake sturgeon are needed to provide information on interpopulation genetic differences. Until this information is available, transplantation of lake sturgeons into other watersheds could cause more harm than good. Indiscriminate transplanting of rainbow trout *Salmo gairdneri* was shown to cause a loss of genetic integrity in local stocks and was presumed to be a factor in the decline of native cutthroat trout *Salmo clarki* populations in certain areas (Nicola 1976; Busack and Gall 1981). With lake sturgeon stocks being threatened throughout much of their range (Miller 1972), any transplanting of this species should be done with extreme care.

Acknowledgments

I thank Greg Kornely and Tina Oman who participated in the field work and aided in the preparation of this manuscript. I also appreciate the partial funding of this study by a concerned group of sportsmen called Sturgeon for Tomorrow.

References

Bartley, D. M., G. A. E. Gall, and B. Bentley. 1985. Preliminary description of the genetic structure of white sturgeon, *Acipenser transmontanus*, in the Pacific Northwest. Pages 105–109 *in* F. P. Binkowski and S. I. Doroshov, editors. North American sturgeons: biology and aquaculture potential. Dr. W. Junk, Dordrecht, The Netherlands.

Becker, G. C. 1983. Fishes of Wisconsin. University of Wisconsin Press, Madison.

Busack, C. A., and G. A. E. Gall. 1981. Introgressive hybridization in populations of Paiute cutthroat trout (*Salmo clarki seleniris*). Canadian Journal of Fisheries and Aquatic Sciences 38:939–951.

Carlson, H., L. M. Andrews, and C. W. Threinen. 1975. Surface water resources of Marinette County. Wisconsin Department of Natural Resources, Madison.

Czeskleba, D. G., S. AveLallemant, and T. F. Thuemler. 1985. Artificial spawning and rearing of lake sturgeon, *Acipenser fulvescens*, in Wild Rose State Fish Hatchery, Wisconsin, 1982–1983. Pages 79–85 *in* F. P. Binkowski and S. I. Doroshov, editors. North American sturgeons: biology and aquaculture potential. Dr. W. Junk, Dordrecht, The Netherlands.

Folz, D. J., D. G. Czeskleba, and T. F. Thuemler. 1983. Artificial spawning of lake sturgeon in Wisconsin. Progressive Fish-Culturist 45:231–233.

Folz, D. J., and L. S. Meyers. 1985. Management of the lake sturgeon, *Acipenser fulvescens*, population in the Lake Winnebago system, Wisconsin. Pages 135–146 *in* F. P. Binkowski and S. I. Doroshov, editors. North American sturgeons: biology and aquaculture potential. Dr. W. Junk, Dordrecht, The Netherlands.

Holmstrom, B. K. 1980. Low flow characteristics of streams in the Menominee–Oconto–Peshtigo River basin, Wisconsin. U.S. Geological Survey, Open File Report, Madison, Wisconsin.

Les, B. L. 1979. The vanishing wild—Wisconsin's endangered wildlife and its habitat. Wisconsin Department of Natural Resources, Madison.

Miller, A. A. 1972. Threatened freshwater fishes of the United States. Transactions of the American Fisheries Society 101:239–252.

Nicola, S. J. 1976. Fishing in western national parks—tradition in jeopardy? Fisheries (Bethesda) 1(6):18–21.

Priegel, G. R. 1973. Lake sturgeon management on the Menominee River. Wisconsin Department of Natural Resources Technical Bulletin 67.

Priegel, G. R., and T. L. Wirth. 1974. The lake sturgeon, its life history, ecology and management. Wisconsin Department of Natural Resources, Publication 4-3600(74), Madison.

Priegel, G. R., and T. L. Wirth. 1975. Lake sturgeon harvest, growth and recruitment in Lake Winnebago, Wisconsin. Wisconsin Department of Natural Resources Technical Bulletin 83.

Rounsefell, G. A. 1958. Anadromy in North American Salmonidae. U.S. Fish and Wildlife Service Fishery Bulletin 58:171–185.

Scott, W. B., and E. J. Crossman. 1973. Freshwater fishes of Canada. Fisheries Research Board of Canada Bulletin 184.

Threader, R. W., and C. S. Brousseau. 1986. Biology and management of the lake sturgeon in the Moose River, Ontario. North American Journal of Fisheries Management 6:383–390.

Thuemler, T. F. 1985. The lake sturgeon, *Acipenser fulvescens*, in the Menominee River, Wisconsin–Michigan. Pages 73–78 *in* F. P. Binkowski and S. I. Doroshov, editors. North American sturgeons: biology and aquaculture potential. Dr. W. Junk, Dordrecht, The Netherlands.

American Fisheries Society Symposium 5:110–122, 1988

Spawning and Early Life History of Lake Sturgeon in the Lake Winnebago System, Wisconsin

JAMES J. KEMPINGER

Wisconsin Department of Natural Resources, Oshkosh, Wisconsin 54901, USA

Abstract.—Populations of lake sturgeon *Acipenser fulvescens* in the Lake Winnebago system are self-sustaining. This phenomenon provided the opportunity to study spawning behavior and the early life history of this unique species. Mature adults migrated as far as 201 km upstream from Lake Winnebago to spawn in the Wolf River. Most spawning occurred in April when water temperatures were 8.5–14.5°C. Adhesive eggs attached to the bottom substrate and hatched from 8 to 14 d after fertilization. Specially designed 950-μm-mesh drift nets were used to determine the extent and timing of downstream movement of larval lake sturgeons. Between 8 and 898 larval lake sturgeons ranging in size from 8 to 22 mm total length were captured annually (1981–1984). Rate of capture ranged from 0.09 to 1.60 individuals/h·m^2 of net opening. Maximum downstream movement occurred at 9–10 d posthatch. Peak drift occurred at night when 6.5 times more larval lake sturgeons were captured than in daytime.

While most populations of lake sturgeon *Acipenser fulvescens* in North America have diminished, the stocks in the Lake Winnebago system are self-sustaining and provide a unique winter spear fishery. Each spring large numbers of mature lake sturgeons are observed at the Shawano Dam and represent probably the largest concentration to occur at any one time and place in North America. This abundant spawning population provided an opportunity to study the early life history of this species.

Interest in the lake sturgeon results from its U.S. federal status as a potentially endangered species, its curiosity value as a living fossil, and its large size, late maturity, and exceptionally long life. The harvest, growth, population structure, and recruitment of adult lake sturgeons in the Winnebago system have been studied since the 1940s (Schneberger and Woodbury 1946; Probst and Cooper 1955; Priegel and Wirth 1974, 1975, 1978; Folz and Meyers 1985). These studies resulted in the enactment of restrictive harvest regulations to protect the population. As nonharvest pressures (pollution, land development, dams) on the resource increase, the need for more information on the biology of the lake sturgeon also increases. Therefore, it was considered essential to study the habitat requirements of egg and larval stages.

Until recently, we thought that lake sturgeons migrated upstream in spring to spawn after the Wolf River was free of ice but before ice had disappeared from Lake Winnebago and the upper river lakes. During this study, lake sturgeons were observed within 40 km of Shawano Dam in greater numbers in fall than in summer. To test the hypothesis that lake sturgeons stay in the Wolf River over winter, 12 adult male and 12 female lake sturgeons were equipped with radio transmitters in the fall of 1985. These fish stayed within the Wolf River all winter and until after spawning (D. Folz, Wisconsin Department of Natural Resources, personal communication.)

Study of the early life history of the lake sturgeon began in 1981 on the Wolf River at the Shawano Dam, Shawano County. This site was chosen because it is a major spawning ground, easily accessible, and shallow enough to sample. The frequency, length, and intensity of spawning, egg deposition, fertilization, incubation, and hatching were studied; downstream movements of larval lake sturgeons, habitat, growth, development, and food habits were determined.

Study Area

The Lake Winnebago system (Figure 1) drains a 15,366-km^2 watershed within the Great Lakes basin (Holmstrom 1982); Lakes Winnebago, Butte des Morts, Poygan, and Winneconne comprise 671 km^2 of surface waters (D. Fago, Wisconsin Department of Natural Resources, personal communication). Two major rivers, the Fox (314 km) and the Wolf (355 km), and two minor rivers, the Little Wolf (109 km) and the Embarrass (99 km), are the principal streams within the watershed. Water levels of the Lake Winnebago system are regulated by a dam on the Fox River at Neenah–Menasha.

Field work for this study took place from 1981 to 1984 on the Wolf River at Shawano Dam, 201

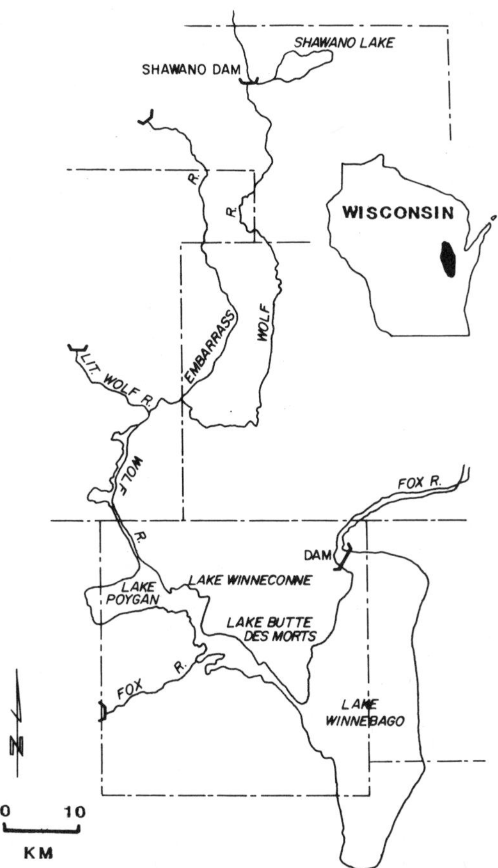

FIGURE 1.—Lake Winnebago (Wisconsin) system study area.

km upstream of Lake Winnebago. Shawano Dam, constructed in 1843, precludes movement of lake sturgeon further upstream. Stream elevation drops 13.6 m from Shawano Dam (237.4 m above sea level) to Lake Winnebago. The steepest drop is within Shawano County where the gradient is 0.19 m/km over the 38-km stretch. The mean width and depth of the river in the study area are 79 m and 1 m, respectively. The substrate in this area of the river is primarily rubble, sand, and boulders. Wolf River water is lightly brown in color; it has a total alkalinity of 97 mg/L as $CaCO_3$ and a conductivity of 219 μS/cm (Andrews and Threinen 1968). The daily average annual flow rate for April and May 1981–1984 at Keshena Falls, 17.6 km upstream of Shawano Dam (2,750 km^2 of watershed), ranged from 32.5 m^3/s to 48.4 m^3/s (U.S. Geological Survey 1981–1984). The Wolf River shoreline in Shawano County is mostly undeveloped land.

Methods

Spawning and egg incubation.—We recorded frequency, intensity, length, and dates of spawning. In early April, a scuba diver counted the number of mature lake sturgeons and observed their behavior within 105 m of Shawano Dam. Throughout the spring, the dates when lake sturgeons were seen surfacing in the middle of the stream (prespawning behavior) were recorded. Fish management personnel, studying movements of lake sturgeons within the Lake Winnebago system, captured mature fish with dip nets and measured, sexed, and tagged them. At the peak of spawning, when eggs flowed freely, eggs were removed and artificially inseminated. A scuba diver placed galvanized metal trays, to be used for egg incubation, on the bottom of the Wolf River between Shawano Dam and the Chicago and Northwestern railroad trestle. The dates of spawning and day of maximum egg deposition were determined. After ice melted from the Wolf River, 10 numbered 1-m^2 trays with 2.5-cm-high sides were placed at depths of 1.8–3.6 m in various habitats with varying flow rates. The diver lifted the trays periodically to monitor egg deposition and survival.

Hatching dates were determined by examination of naturally and artificially inseminated eggs. Females were spawned and the eggs broadcast over stones (≤3.8 cm in diameter) placed in a 0.5-m^2 wash tub with water just covering the stones. In 1981, Monsanto AstroTurf was also used. After egg deposition in the tub, males were induced to spawn, thus fertilizing the eggs. The stones, with eggs adhering to them, were removed; the eggs were counted and placed in 0.3-m^2 galvanized trays with 2.5-cm sides. Eight of the 16 trays were covered with a 1,700-μm-mesh screen to exclude predators, and the other 8 were left uncovered. The trays were placed in the river by a scuba diver in approximately 1.5-m-deep water. They were lifted periodically to monitor survival.

Cumulative daily water temperature units (CTU) were calculated to determine the approximate hatching date by means of the equation,

$$\text{CTU} = \sum_{i=1}^{n}(x_i - K);$$

K = the constant, 5.8°C;
x_i = mean daily water temperature (°C) for day i calculated from readings taken at 4-h intervals;
n = number of days from peak spawning to day of hatching.

Negative values of $x_i - K$ were not included in the sum. This equation differs from that used by Piper et al. (1982) in that it includes a correction factor, K, determined from the observed values. This was done by calculating CTU from the equation for all values of K between 2 and 10°C at 0.1°C increments and searching for that value of K that gave the minimum between-year variance, as well as the minimum range and minimum sum of deviations from median CTU. The final equation should result in a nearly constant CTU for all years and thus provides a more reliable prediction of hatching dates.

Larval emergence and development.—On the day of peak spawning, drift nets were set to collect eggs, yolk-sac larvae, and larvae. The nets consisted of rectangular (0.43 × 0.86 m, area = 0.37 m^2) or circular (0.91-cm diameter, area = 0.66 m^2) frames that supported a 3-m-long, 950-μm-mesh bag net with a removable cup at the end of the net. In 1981, two rectangular nets and one circular net were set. The nets were sampled for 1 h at dawn, 0800, 1200, and 1600 hours, dusk, and 2400 hours. Three rectangular frames and one circular frame were used in 1982. Sampling was scheduled in shifts (0700 to 1500, 1500 to 2300, and 2300 to 0700 hours) with a pattern of three shifts on and one shift off, then three shifts on and two shifts off within each 3-d period. In 1983 and 1984, two rectangular nets were set singly and two were stacked to sample the water column from the bottom to near the surface. Nets were sampled hourly between 2000 and 0500 hours in 1982, 1983, and 1984. Larval lake sturgeons were collected during 178 net-hours of sampling (29 April–15 May) in 1981, 1,219 net-hours (7–26 May) in 1982, 900 net-hours (6 May–3 June) in 1983, and 683 net-hours (11 May–1 June) in 1984.

Drift nets were placed in waist-deep water, on gravel and rock substrate, 150 m downstream of the Shawano Dam. Samples collected in drift nets were examined in an enamel tray, and all larval lake sturgeons were sorted and counted. Catch per unit of effort was the number of larval lake sturgeons collected per hour per square meter of net opening. For data analysis, the sampling day began at 2000 hours the previous day; for example, the 27 April sampling day extended from 2000 hours on 26 April to 1959 hours on 27 April. Specimens collected were fixed in 10% formalin and 2 weeks later were transferred to 60% isopropyl alcohol. All larvae were measured to the nearest millimeter total length (TL). The classification of larval lake sturgeon by size in this report is taken from Auer (1982).

Of the 898 larval lake sturgeons (11–21 mm TL) captured in 1982, 231 were examined to describe developmental morphology. Larvae without visible yolk sacs were dissected, and their stomach contents were examined under a dissecting microscope.

Environmental variables.—Physical parameters were measured in the Wolf River at the Shawano boat landing. A staff gage calibrated in 3.2-mm increments was permanently installed. Water levels and atmospheric conditions (occurrence of precipitation, wind velocity, and cloud cover) were recorded hourly on sampling days. A Taylor weekly thermograph was installed during ice-free periods to continuously record air and water temperatures. Daily precipitation records for the City of Shawano were obtained from U.S. National Climatic Center (1981–1984). Water velocity was measured with a Marsh-McBurney flow meter at the surface, middepth, and bottom over various spawning habitats between the Shawano Dam and the Chicago and Northwestern railroad trestle.

Results

Spawning and Egg Incubation

During all 4 years of this study, adult lake sturgeons were observed in early April by a scuba diver at the Shawano Dam. Fish were inactive but wary when approached by the diver. As the spring season progressed, lake sturgeons surfaced in swift current, always avoiding the shoreline. With a rise in water temperature, large numbers of lake sturgeons were observed spawning over boulders in rapid current at the foot of the dam. Spawning fish preferred a shoreline with relatively strong (more than 15 cm/s) currents and shallow water. A diver could swim among the groups of spawning lake sturgeons and touch a fish without disturbing the school.

During mating and spawning, males swim above and alongside a female and in the same direction, usually against the current. Six to eight male lake sturgeons accompany one female. When spawning takes place, eggs are extruded for a short period and males vibrate simultaneously alongside a female. Ripe females taken from the Wolf River during the peak of spawning have eggs that run freely for 7–10 s. If a female is held in a tank for 10–15 min after oviposition, the eggs will again run freely. After a spawning foray, which can consist of several acts, the group drifts downstream or swims into deeper water. Male spawners outnumber females because they mature at 14–16 years of age, whereas the females mature at age 24–26 years (Priegel and Wirth 1974). Females spawn every 4–6 years (E. N. Magnin, University of Montreal, and G. R. Priegel, Wisconsin Department of Natural Resources, unpublished data) while males usually spawn every other year.

Each year we noted at least two temperature-dependent spawning periods, with the most intensive spawning (peak) occurring during the first period. When a cold front passed through during spawning, a slight drop in water temperature would cause the lake sturgeon to disperse. As temperature rose, spawning activity resumed.

In 1981, the spawning peak occurred between 14 and 17 April, when the water temperature was 8.3–11.1°C (Table 1; Figure 2). A drop in air temperature on 18 April lowered the water temperature by 1.5–3.0°C, and spawning ceased. Lake sturgeons returned to the shoreline to spawn on 5 May, when the water temperature reached 13.0°C.

In 1982, water temperature steadily increased from 6.1°C on 23 April to 10.5°C on 30 April, the day of peak spawning. Water temperature increased to 15.5°C by 6 May, at which time a cold front reduced water temperature to 11.6°C, and spawning ceased. A second spawning took place on 12 May 1982 when the water temperature reached 15°C.

In 1983, water temperature increased from 4.4°C on 19 April to 13°C on 26 April. Peak spawning occurred between 1800 and 2300 hours on 28 April when the water temperature was 14.2°C. Thereafter a cold front lowered the water temperture to 9°C on 8 May, and spawning ceased. A second spawning took place on 13 May when the water temperature was 13.9°C.

TABLE 1.—Water temperatures during spawning by lake sturgeon in Wolf River, Wisconsin, 1981–1984.

	Peak spawning		Secondary spawning	
Year	Date	Water temperature (°C)	Date	Water temperature (°C)
1981	14–17 Apr	8.3–11.1	5–7 May	13.0–13.3
1982	30 Apr–6 May	10.5–16.1	12 May	15.0–15.8
1983	28 Apr	13.6–14.7	13 May	13.9–14.4
1984	28 Apr	12.8–14.7	18–19 May	16.1–18.9
			8 Jun	21.1–23.3

In 1984, water temperatures warmed quickly from 7°C on 23 April to 14.7°C on 27 April. On 27 April, lake sturgeon were easily captured along the shoreline with dip nets. Two females captured that day did not extrude eggs when their abdomens were pressed, but ovulating females were observed 1 d later when the water temperature was 13.3°C. By 30 April, a cold front had passed through the area, dropping the water temperature to 9.0°C and dispersing the spawning sturgeon. Fish were observed again surfacing at the dam on 18 and 19 May when the water temperature was 16.1°C. A few sturgeon were seen surfacing at the Shawano Dam on 8 June, when the water temperature was 21.1°C.

During the spawning act, eggs are randomly scattered and adhere to rocks and any other objects in the water. Eggs were abundant along most of the shoreline between the Shawano Dam and Chicago and Northwestern railroad trestle during all 4 years of the study. The scuba diver found lake sturgeon eggs scattered from the shoreline to a depth of 2.5 m at the Shawano Dam. Eggs extruded in the swift water of the chutes immediately downstream of Shawano Dam were most abundant, with egg clumps as much as 15 cm thick. Eggs were also abundant within a meter of shore where the water velocity was greatest.

Of all the galvanized trays set as spawning substrates, eggs were found only at those stations where the water velocity exceeded 10 cm/s. The mean number of eggs deposited on the 10 1-m^2 trays placed between the Shawano Dam and the Chicago and Northwestern railroad trestle on the day of maximum deposition was 1,035 (range 0–9,600) in 1982, 8,805 (range 0–71,140) in 1983, and 6,806 (range 0–31,820) in 1984. One 1-m^2 tray set 10.5 m downstream of the hydroelectric plant on 29 April 1983 contained 71,140 eggs. When the diver lifted the tray, an estimated 75% of the eggs sloughed off and washed downstream; this tray may have contained as many as 250,000 eggs.

The scuba diver observed crayfish *Orconectes* spp., mud puppies *Necturus maculosus*, redhorse *Moxostoma* spp., and common carp *Cyprinus carpio* feeding over lake sturgeon egg deposits. Adult postspawning lake sturgeons also appeared to be consuming deposited eggs. Fungus overgrowth of clumped egg masses was observed in all 4 years of the study.

Egg mortality could also be caused by sediment deposition and water level fluctuation. On 24 April 1981, increased water discharge from the Shawano Dam raised the river level 32 cm be-

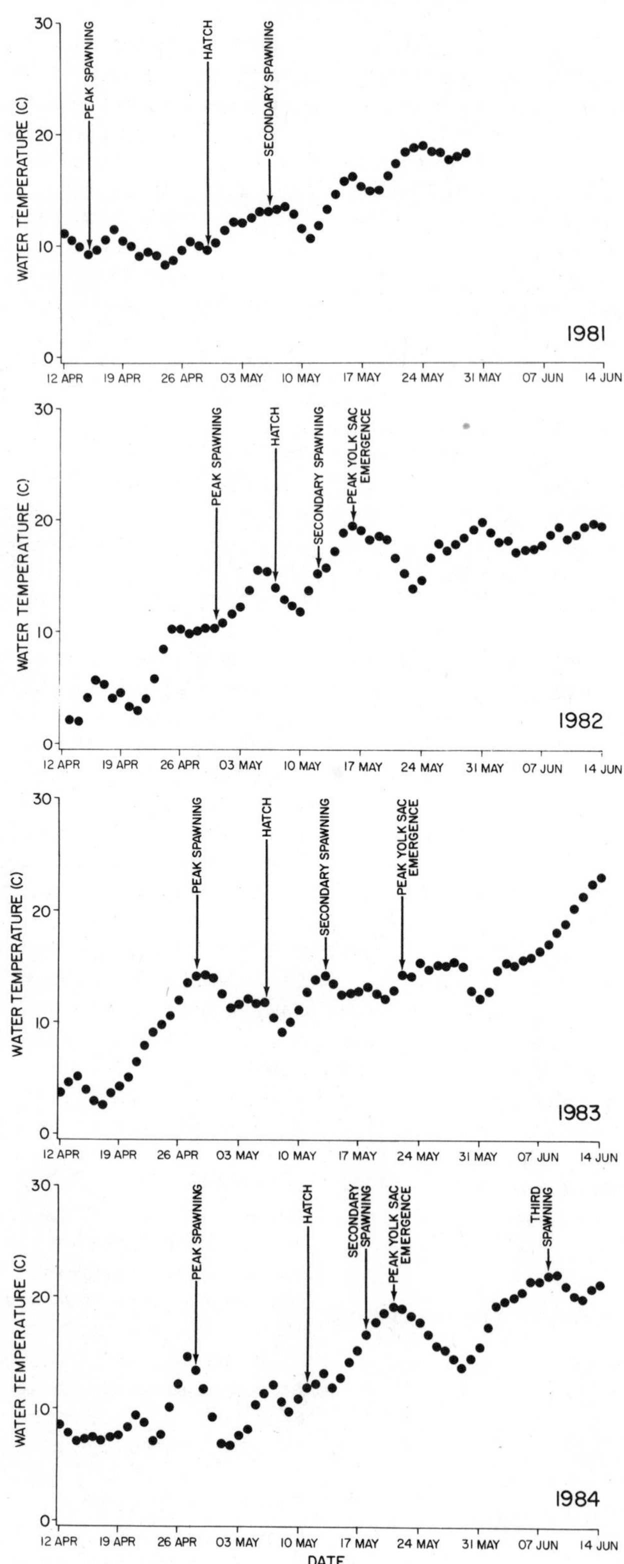

FIGURE 2.—Wolf River water temperatures in relation to dates of peak spawning, hatch, and peak emergence of lake sturgeons, 1981–1984.

tween 0800 hours and midnight, dislodging many eggs. By the next morning the level was lowered 12 cm below that at spawning, and many eggs were exposed and desiccated.

Mortality rates of sturgeon eggs deposited on trays were high. In 1982, we observed a 42% reduction in the mean number of eggs per tray from 2 May (maximum deposition) to 4 May (1,035 and 601 eggs/tray, respectively). On 7 May the embryos were hatching, and many empty egg cases were found on two of the trays.

In 1983, the number of eggs was reduced 97% from a mean of 8,805 eggs/tray on the day of maximum deposition, 29 April, to 251 eggs on 2 May. There was a further reduction of 83% from 2 to 6 May (43 eggs). On 9 May, hatching was in progress and tails of yolk sac larvae were observed emerging from the egg cases.

In 1984, heavy rains and increased current velocity hampered the diver observations. On 30 April, 2 d after maximum egg deposition, only one tray was examined. Less than 1% of the initially deposited 31,820 eggs remained on that tray. On 3 May we were able to check seven trays which contained only 2.8% of the initially deposited eggs (mean, 6,806).

Survival of the artificially fertilized eggs attached to the substrate (stones) and placed on protected 0.3-m^2 screen-covered trays was better than on uncovered trays. Average survival per tray in uncovered trays was 1.4% (range, 0–5.1%). Survival on trays covered with screens averaged 8.9% (range, 7.8–9.9%).

Silt and sand sedimentation was observed on the trays in 2 of the 4 years of the study. In 1983, a fine silt, transported from a private construction site, was found on the trays. There was no indication of increased egg mortality. The installation of a municipal storm sewer, located on the east bank upstream of the Chicago and Northwestern railroad trestle, caused the heavy sedimentation of sand in 1984; eggs on the trays and those on the spawning ground were completely covered. Egg mortality on these trays was 100%.

In 1981, the incubation period was 14 d (Table 2). Eggs fertilized on 0.3-m^2 trays on 15 April were observed hatching on 29 April. Cold weather after peak spawning (14–17 April) lowered the water temperature from 11.4°C on 18 April to 8.3°C by 24 April (Figure 2). The CTU during the 15–29 April incubation period was 54.9. In 1982, the eggs deposited and fertilized on 30 April were hatching on 7 May. The incubation period was 8 d, and the CTU was 57.7 (Table 2). In 1983, the peak of egg deposition occurred on 28 April. Embryos began hatching on trays on 6 May, an the CTU from 28 April to 6 May was 57.9. In 1984, eggs fertilized on 28 April on 0.3-m^2 trays were in the process of hatching on 11 May; this amounted to an incubation period of 14 d and a CTU of 59.9.

Larval Emergence and Development

Larval lake sturgeons were captured annually (1981–1984) 150 m downstream of the Shawano Dam. Sampling with three or four drift nets, resulted in the capture of 1,232 larvae ranging from 8 to 22 mm TL. Annual catch in the drift nets ranged from 8 larvae in 1981 to 898 in 1982. Catch per unit of effort ranged from 0.09 to 1.60 larvae/h·m^2 of net opening (Table 3). Daily water temperatures and dates of spawning, hatching, and peak larval emergence for years 1981–1984 are shown in Figure 2.

In 1981, the first yolk-sac larva (13.1 mm TL) was captured at 0100 hours on 4 May. Embryos on the 0.3-m^2 trays began hatching on 29 April. Eight yolk-sac larvae, 13.1–19.0 mm TL, were captured between 29 April and 15 May. In 178 h of sampling (Table 3), larval lake sturgeons were captured at the rate of 0.09/h·m2.

In 1982, yolk-sac larvae (9–11 mm TL) were captured in drift nets at noon on 7 May. A 10-mm TL yolk-sac larva, captured that first day of emergence, is illustrated in Figure 3. In 1,219 h of sampling, 898 larval lake sturgeons were captured between 7 and 26 May at the rate of 1.60/h·m^2 (Table 3). Only 79 (8.8%) of the yolk-sac larvae were captured in the first 8 d posthatch (Figure 4).

TABLE 2.—Incubation period and cumulative temperature unit (CTU) to day of hatch for lake sturgeon eggs incubated on trays in Wolf River, Wisconsin, 1981–1984.

Year	Incubation period (d)	CTU[a]
1981	14	54.9
1982	8	57.7
1983	9	57.9
1984	14	59.9

[a]CTU = Σ (mean daily water temperature − 5.8°C).

TABLE 3.—Drift-net catches and catch rates of larval lake sturgeons in Wolf River, Wisconsin, 1981–1984.

Year	Sampling: Date	Sampling: Hours	Catch (number)	Catch rate (number/h·m^2 of net opening)
1981	29 Apr–15 May	178	8	0.09
1982	7–26 May	1,219	898	1.60
1983	6 May–3 Jun	900	261	0.78
1984	11 May–1 Jun	683	65	0.26

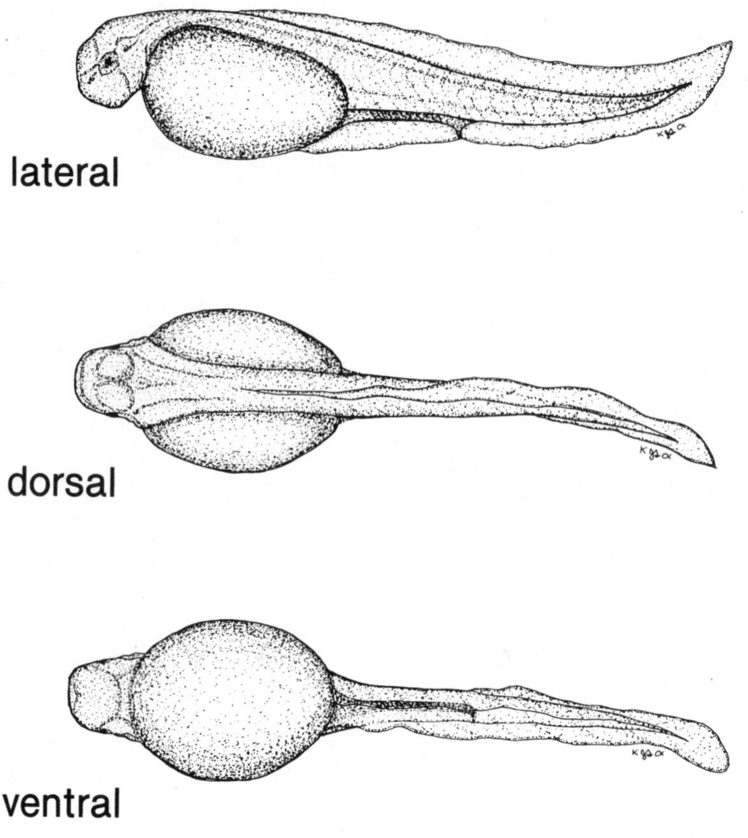

FIGURE 3.—Yolk-sac larva of lake sturgeon captured 1 d posthatch (10 mm total length), 1982.

Yolk-sac larvae (12–14 mm TL) captured 9–11 May had faintly pigmented skin and eyes; the yolk sac was present, and the spiral valve was evident. The mouth had a slight inward fold, barbels were evident, and gills, fins, and lateral lines were undeveloped. At 15 mm TL, a pigmented melanin plug appeared in the spiral valve (Wang et al. 1985).

Larval drift was greatest between the 8th and 14th days posthatch (16–22 mm) when 76% of the larval fish were captured (Figure 4). Peak movement (catch rate, 6.5 larvae/h·m^2 of net opening) in 1982 was on the 9th and 10th days (16–17 May) posthatch; 39% of the 898 yolk-sac larvae captured were captured at that time.

The mean size of yolk-sac larvae at peak movement was 17.5 mm TL. The body and eyes were pigmented, the yolk sac was partially absorbed, and the melanin plug in the spiral valve was conspicuous. Barbels, mouth, gills, fins, and the lateral line were well developed (Figure 5). On the 14th–20th days posthatch, 78 (8.5%) larval lake sturgeons (19–22 mm) were captured. The melanin plug in the spiral valve was still evident.

Night samples (taken between 2100–0500 hours) constituted one-third of the sampling effort in 1982. Between 7 and 26 May, 686 larval lake sturgeons (76%) were captured during darkness. The rate of capture at night from the 8th through the 14th days posthatch was 8.9 larvae/h·m^2 compared to 1.0/h·m^2 during the daylight sampling period (Figure 6).

When these data were summarized by hour, irrespective of date, an average of 4.2 larval lake sturgeons/h·m^2 were captured between 2100 and 0200 hours (Figure 7). Between 0500 and 2100 hours, 1.2 larvae/h·m^2 or fewer were captured. The fewest larvae (0.10/h·m^2) were captured between 1700 and 1800 hours. Peak daily and hourly movements were between midnight and 0500 hours on 15 May when 188 yolk-sac larvae were captured at the rate of 25.6/h·m^2.

In 1983, the first six yolk-sac larvae were captured on 5–6 May. These were abnormal, appear-

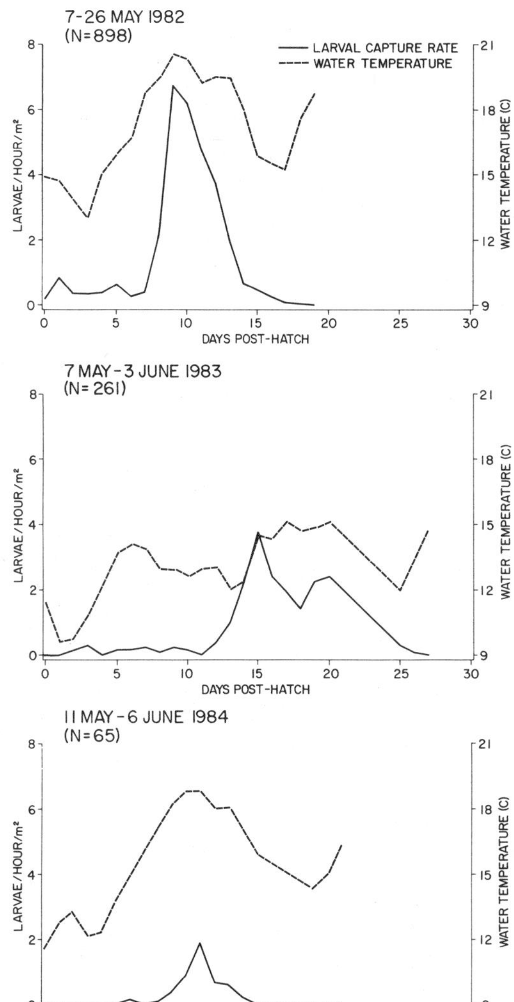

FIGURE 4.—Wolf River water temperatures (°C) and mean hourly catch rates (number/h·m^2 of net opening) of larval lake sturgeons in relation to days posthatch, 1982–1984; N = total number of captured larvae.

ing to be undeveloped and to have prematurely emerged from their eggs. The water temperature dropped 4.5°C between the day of peak spawning (28 April) and the day after embryos began hatching on 7 May (Figure 2). The first normal yolk-sac larva (11 mm TL) was captured 2 d posthatch at 2000 hours on 8 May. In 900 net-hours of sampling (Table 3), 261 larval lake sturgeons were captured between 6 May and 3 June at a mean rate of 0.78/h·m^2. Peak larval movement in 1983 occurred between 2100 and 0500 hours on 21 May or 15 d posthatch when larvae were captured at a rate of 3.7/h·m^2 (Figure 4). The mean size of yolk-sac larvae at peak drift was 18.4 mm TL.

The seasonal and hourly peak of capture was between 0300 and 0500 hours on 22 May when 16 yolk-sac larvae were captured at the rate of 5.40/h·m^2. The lower of the two stacked drift nets captured 1.08 compared to 0.77/h·m^2 captured in the upper drift net. When catch rates were summarized by time irrespective of date, most larval fish (1.32/h·m^2) were captured between 2200 and 2300 hours. Hourly catch rates within the 2000–0500-hours sampling schedule for 1983 ranged from 0.51–0.95/h·m^2.

In 1984, 683 hours of fishing between 11 May and 1 June captured 65 larval lake sturgeons (13–21 mm TL) at the rate of 0.26/h·m^2 (Table 3). The first yolk-sac larvae (13 and 14 mm), approximately 7 d posthatch from the first hatch, were captured at 1600 hours on 17 May. The peak of larval emergence from the first hatch occurred on 21 May (1.9/h·m^2) or 11-d posthatch when the mean size of yolk-sac larvae was 18.8 mm TL (Figure 4). Two 16-mm yolk-sac larvae were captured from the second hatch on 1 June.

The seasonal and hourly peak of capture occurred between 2100 and 2400 hours on 22 May when 18 yolk-sac larvae were captured at the rate of 4.05/h·m^2. The lower of the two stacked drift nets captured 0.25 in contrast to 0.38/h·m^2 in the upper net. When larval catch rates were summarized by time irrespective of date, most larval lake sturgeons (0.57/h·m^2) were captured between 2300 and 2400 hours. Hourly catch rates during the 1984 sampling season between 2000 and 0500 hours ranged from 0.11 to 0.39/h·m^2.

Only one larva (21 mm) of the 845 examined in 1982 for food content had begun exogenous feeding; it consumed one chironomid larva. One 20-mm lake sturgeon larva captured in 1983 had consumed one copepod nauplius.

In another segment of this study, we sampled with drift nets at the county highway CCC bridge 12.8 km downstream of Shawano boat landing. Here we captured two (30 and 31 mm TL and 0.1 and 0.12 g, respectively) lake sturgeon larvae on 2 June 1982. The 30-mm larva had consumed one Baetidae nymph, and the 31-mm larva contained six Baetidae nymphs and two chironomid larvae.

The relationship between total length of larvae and their age (days posthatch) varied among years (Figure 8). Growth in length appeared to be approximately linear up to a length of 20 mm. There were statistically significant differences between the regression line slopes for 1982, 1983, and 1984

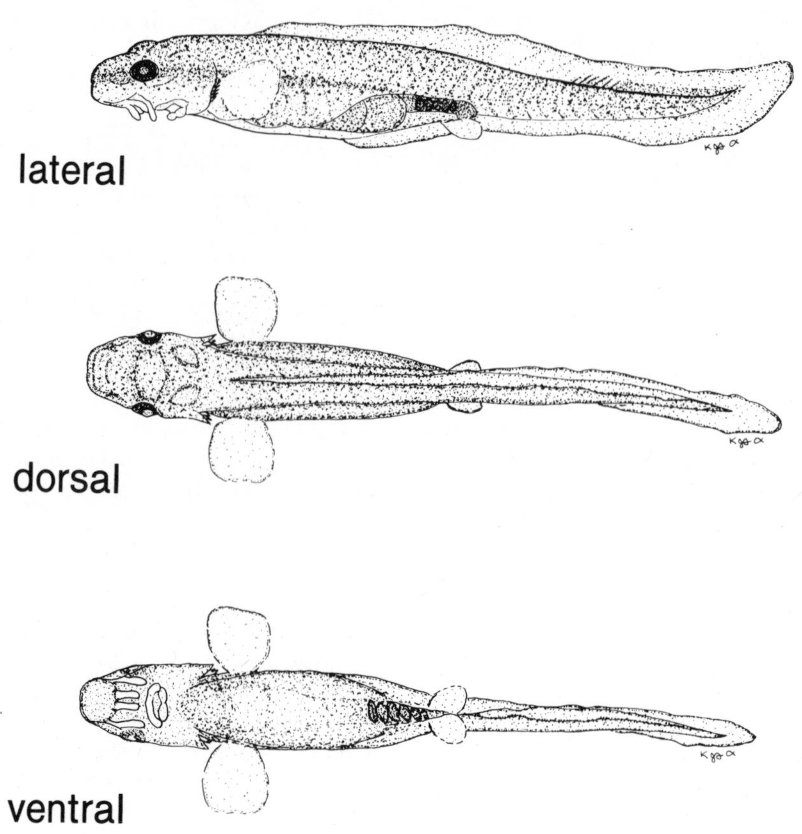

FIGURE 5.—Yolk-sac larval lake sturgeon (17.5 mm total length) captured 10 d posthatch, 1982.

($P < 0.001$), indicating slower growth in 1983, when mean daily water temperature was cooler than in 1982 and 1984.

Discussion

Spawning and Egg Incubation

Water temperature strongly influenced the spawning of lake sturgeon. Although spawning took place over a wide range of temperatures (8.3–23.3°C), the peak generally occurred between 10 and 14°C (range, 8.3–14.2°C). Slight decreases in water temperature (as little as 1.5–3.0°C) resulted in the cesstion of spawning.

The best year in terms of reproductive success was 1982. Water warmed gradually over a period of several days before and during spawning and several days after the day of peak spawning (Figure 2). In the other 3 years, cessation of spawning due to drop in temperature occurred within 1 to 3 d after peak spawning.

The rate of water-temperature increase over the range 8–15°C appears to be important for the onset of spawning activity. In contrast to the gradual warming that occurred in 1982, a sudden warming of the water in 1984 (from 7.0 to 14.7°C within 3 d) did not evoke immediate spawning. One day later, however, when a cold front came through and the water temperature dropped to 13.3°C, spawning took place. The inducement to spawn, then, appears to be a combination of suitable water temperatures and some minimum period of time required for spawning response.

In spite of some negative effect of temperature fluctuation, lake sturgeon spawning in 1981–1984 was of a magnitude that provided abundant egg deposition annually at the Shawano Dam. Eggs were most abundant just below the dam in the swiftest currents; no eggs were found in the areas where the current velocity was less than 10 cm/s. Harkness and Dymond (1961) reported that 11,000 eggs/kg of fish are produced by an average lake

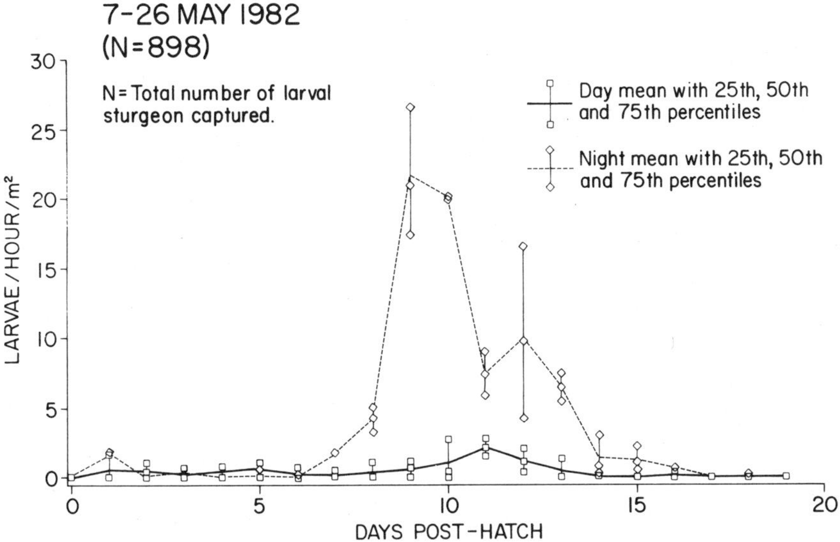

FIGURE 6.—Day and night mean catch rates of the larval lake sturgeons in the Wolf River at different ages posthatch, 1982. Vertical bars denote 25, 50, and 75 percentiles.

sturgeon female. A 25-year-old female, in its first year of maturity, 1.4 m long and weighing 22.5 kg, has the potential of producing 250,000 eggs.

The quantity of eggs alone may not be positively correlated with reproductive success. In 1982, the best year in terms of hatching success as indicated by drift-net captures, egg deposition on the 1-m^2 trays was, on the average, only 12% of that in 1983 and 20% of 1984.

When lake sturgeon embryos were exposed to fluctuating or low water temperatures, the result was a longer incubation period that probably increased their exposure to bacterial and fungal diseases as well as to predation. Data from the 1-m^2 trays suggest that egg survival in 1982, the year with the shortest incubation period, was better than in 1983 or 1984. A comparison of egg mortalities on covered and uncovered 0.3-m^2

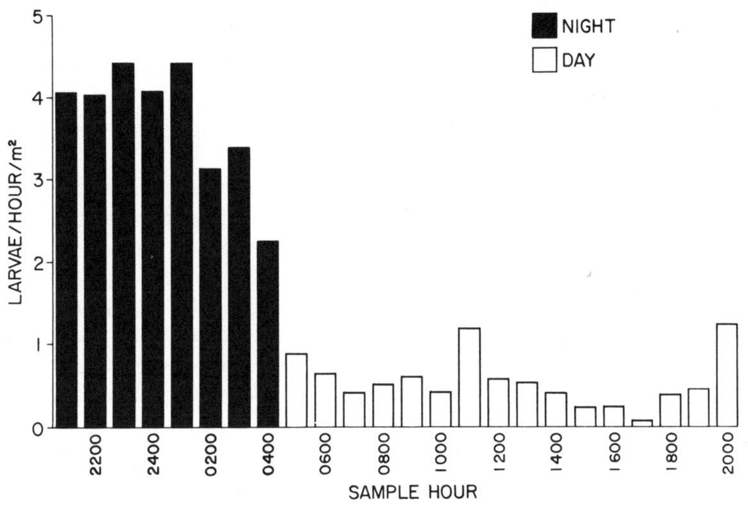

FIGURE 7.—Mean hourly day and night catch rates of larval lake sturgeon from the Wolf River, 1982.

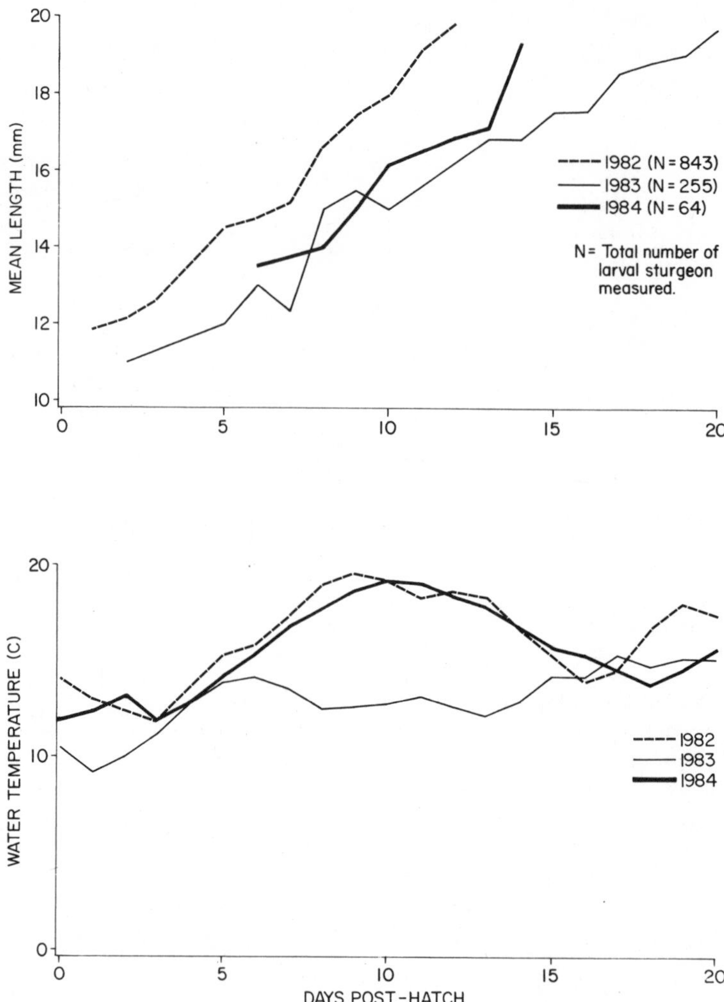

FIGURE 8.—Growth rates of lake sturgeon larvae and observed water temperatures of the Wolf River, 1982–1984.

trays suggests that predation was one likely source of mortality, but major mortality was due to some other causes, possibly fungal or bacterial disease. Although most of the observed egg clumps were covered with fungus, excessive sand or silt sedimentation resulting from local development projects could have produced similar mortality.

There was no indication that lowering or raising the water level affected hatching success at the immediate proximity to the dam, but below the dam many embryos were exposed to air, and they became desiccated. Khoroshko (1972) confirmed increased spawning success of *Acipenser stellatus* in the Volga River, when water discharge was regulated during sturgeon spawning below the dam.

The incubation period for lake sturgeon eggs was temperature dependent and ranged from 8 to 14 d. The CTU values ranged from 54.9 to 59.9 (mean, 57.6) at the time of hatching. This five-unit difference is sufficiently small to have allowed a reasonable prediction of hatch within 1 d. If the 1981 data are ignored, the difference is just slightly over two units. Czeskleba et al. (1985) incubated Fox River lake sturgeon eggs at constant temperatures in 1982 and 1983. When our correction constant (5.8°C) is applied to their data, those eggs that were incubated at 13°C in 1982 began hatching at a CTU of 57.6 (8 d after fertilization). Eggs incubated at 16°C the same year began hatching in 4 d (CTU, 40.8) about 1.5 days earlier than our data would have predicted. In 1983, eggs incubated at 13°C began hatching in 9 d instead of 8. The mean CTU for these three sets of data was 57.7. Wang et al. (1985) found similar variabilities, 66.4 and 53.2 CTUs at 10 and 20°C, respectively, but the mean CTU for the two treatments was 58.8.

Larval Emergence and Development

The abundance of larvae measured by drift-net catch was highest in 1982 of the 4 years studied. The gradual warming of water during the prespawning, spawning, and most of the incubation period resulted in the lowest egg mortality, highest catch rates of larvae, and best growth during 20-d posthatch (Figure 8). These conditions also produced the earliest peak of larval emergence (9–10-d posthatch) of the 4 years.

After their emergence, the yolk-sac larvae of lake sturgeon are nocturnal drifters. The extensive day and night sampling in 1982 demonstrated that nightly catch rates were nine times higher than daytime catch rates during the peak of emergence (Figure 6). Newly hatched yolk-sac larvae in the Wild Rose State Fish Hatchery (Wisconsin) swam from the hatching jars and aggregated in the darkest areas of the AstroTurf mats (Czeskleba et al. 1985). Harkness and Dymond (1961) reported that yolk-sac larvae of lake sturgeon concentrated in dark corners or under rocks placed in the aquarium.

The principal bottom materials at the spawning site below the Shawano Dam are cinders, which have many crevices. Historically, cinders were deposited in the Wolf River at the Shawano Dam when steam-operated railroad locomotives were in use. The dense substrate of cinders at the spawning site should be an ideal habitat for newly hatched lake sturgeon yolk-sac larvae.

Acknowledgments

Special recognition is given to T. F. Rasman for providing his scuba diving expertise to observe spawning lake sturgeons and for setting and monitoring the sampling gear. A. M. Forbes and H. E. Snow helped design the study, and D. J. Folz familiarized me with the study area. Many dedicated Department of Natural Resource employees under the guidance of the late J. E. Keppler collected the field data. Statistical advice was provided by E. L. Lange and P. W. Rasmussen. Drawings and graphs were produced by K. Scheidegger and R. G. Burton, respectively. T. D. Pellett contributed to data analysis and writing, and L. M. Christenson reviewed the manuscript. K. E. Scott typed and revised the manuscript. Support for this study was provided, in part, by the Federal Aid in Fish Restoration Act, Project F-83-R. Sturgeon for Tomorrow Incorporated, a nonprofit sportsmen's group, contributed money toward the purchase of equipment used in this study.

References

Andrews, L. M., and C. W. Threinen. 1968. Surface water resources of Shawano County. Wisconsin Department of Natural Resources, Madison.

Auer, N. A., editor. 1982. Identification of larval fishes of the Great Lakes basin with emphasis on the Lake Michigan drainage. Great Lakes Fishery Commission, Special Publication 82-3, Ann Arbor, Michigan.

Czeskleba, D. G., S. AveLallemont, and T. F. Thuemler. 1985. Artificial spawning and rearing of lake sturgeon, *Acipenser fulvescens*, in Wild Rose State Fish Hatchery, Wisconsin, 1982–1983. Pages 79–85 *in* F. P. Binkowski and S. I. Doroshov, editors. North American sturgeons: biology and aquaculture potential. Dr. W. Junk, Dordrecht, The Netherlands.

Folz, D. J., and L. S. Meyers. 1985. Management of the lake sturgeon, *Acipenser fulvescens*, population in the Lake Winnebago system, Wisconsin. Pages 135–146 *in* F. P. Binkowski and S. I. Doroshov, editors. North American sturgeons: biology and aquaculture potential. Dr. W. Junk, Dordrecht, The Netherlands.

Harkness, W. J. K., and J. R. Dymond. 1961. The lake sturgeon, the history of its fishery and problems of conservation. Ontario Department of Lands and Forests, Toronto.

Holmstrom, B. K. 1982. Drainage area data for Wisconsin streams. U.S. Geological Survey and Wisconsin Department of Transportation Division of Highways, Madison, Wisconsin.

Khoroshko, P. N. 1972. The amount of water in the Volga basin and its effect on the reproduction of sturgeons (*Acipenseridae*) under conditions of regulated and normal discharge. Journal of Ichthyology 12:608–615.

Piper, R. G., I. B. McElwain, L. E. Orme, J. P. McCraren, L. G. Fowler, and J. R. Leonard. 1982. Fish hatchery management. U.S. Fish and Wildlife Service, Washington, D.C.

Priegel, G. R., and T. L. Wirth. 1974. The lake sturgeon, its life history, ecology and management. Wisconsin Department of Natural Resources Publication 4-3600(74).

Priegel, G. R., and T. L. Wirth. 1975. Lake sturgeon harvest, growth and recruitment in Lake Winnebago, Wisconsin. Wisconsin Department of Natural Resources Technical Bulletin 83.

Priegel, G. R., and T. L. Wirth. 1978. Lake sturgeon populations, growth and exploitation in Lakes Poygan, Winneconne and Lake Butte des Morts, Wisconsin. Wisconsin Department of Natural Resources Technical Bulletin 107.

Probst, R. T., and E. L. Cooper. 1955. Age, growth and production of the lake sturgeon (*Acipenser fulvescens*) in the Lake Winnebago region, Wisconsin. Transactions of the American Fisheries Society 84: 207–227.

Schneberger, E., and L. A. Woodbury. 1946. The lake sturgeon *Acipenser fulvescens* Rafinesque, in Lake Winnebago, Wisconsin. Transactions of the Wisconsin Academy of Sciences, Arts and Letters 36: 131–140.

U.S. Geological Survey. 1981–1984. Water resources data. Wisconsin, Water Data Report WI-1981–1984-1.

U.S. National Climatic Center. 1981–1984. Climatological data—Wisconsin, 1981–1984. Asheville, North Carolina.

Wang, Y. L., F. P. Binkowski, and S. I. Doroshov. 1985. Effect of temperature on early development of white and lake sturgeon, *Acipenser transmontanus* and *A. fulvescens*. Pages 43–50 *in* F. P. Binkowski and S. I. Doroshov, editors. North American sturgeons: biology and aquaculture potential. Dr. W. Junk, Dordrecht, The Netherlands.

American Fisheries Society Symposium 5:123–130, 1988

Sampling of Larval European Smelt: A Factorial Experiment

MIKAEL HILDÉN[1]

Department of Limnology, University of Helsinki, Viikki, E-Building
SF-00710 Helsinki, Finland

LAURI URHO

Finnish Game and Fisheries Research Institute, Post Office Box 202
SF-00151 Helsinki, Finland

Abstract.—Samples of larval fish are often highly variable. The variability of catches of larval European smelt *Osmerus eperlanus* was studied in a factorial experiment with four sampling speeds, two sampling depths, and three sampling days. Day and depth effects significantly influenced the number of larvae caught in all length-groups between 6 and 37 mm. Sampler speeds between 1 and 3 m/s generally explained a smaller proportion of the variability, and this variable was not significant for larvae 6–12 mm long. The high day-to-day variability was due to changes of location of the larvae but also to changes in their vertical position in response to weather conditions. A minimum sampler speed of 2 m/s was found adequate for sampling larvae in the size range 6–24 mm with the modified Gulf V high-speed sampler.

Numerous quantitative samplers of fish larvae are used, and some of these have reached the status of standard gear. Among the standard ichthyoplankton samplers are the Gulf high-speed sampler and its modifications (Bridger 1958; Nellen and Schnack 1975). The design has been thoroughly tested, and its hydrodynamic properties are reasonable (Bjørke et al. 1974). The densities of larval fish estimated from catches by this sampler nevertheless vary substantially within fairly restricted sampling areas. The variation may partly be due to net avoidance (Noble 1970; Colton et al. 1980), which depends on, among other things, sampler speed and the size of the aperture of the sampler. In addition, the size and species of larvae influence net avoidance (Thayer et al. 1983). Larval behavior such as vertical migration, aggregation, and movement may also cause variability of catches (Cole and MacMillan 1984).

The variation in the density estimates may mask differences in density between areas and years. In extreme cases, such variability makes larval samples useless for estimates of recruitment, environmental impact assessments, and other purposes.

In this paper, we focus on the sampling problem by analyzing the variation in the catches of larval European smelt *Osmerus eperlanus* within a relatively small inlet in order to study the influences that sampling date, sampling depth, and sampler speed have on density estimates of larvae in four length-groups between 6 and 37 mm.

Methods

Study area.—Larval European smelt were sampled in the polluted inlet Vanhankaupunginlahti of the northern Baltic Sea off Helsinki (60°12′N, 25°00′E). The inlet is shallow (depth rarely exceeds 2 m), and it is an important spawning and nursery area for European smelt. The water is turbid and the Secchi disc transparency varies between 0.2 and 0.5 m. There are practically no macrophytes in the water except for reed belts along the shores.

Sampler.—The sampler was a boat-mounted modification (Hudd et al. 1984) of the Gulf V sampler with an aperture diameter of 0.19 m (Nellen and Schnack 1975). At sampling, two parallel samplers were about 1 m from the bow on each side of the boat to avoid the wash and drag. The flowmeter was placed in the aperture of the sampler. The net had a mesh size of 300 μm and ended in a collecting bottle.

Experimental design and analysis.—The experiment was designed to give replications at all levels of the factors under consideration. In the final design, four sampler speeds were studied at two depths during 3 d. The classes of sampler speed were 0.9–1.1 m/s, 1.6–1.7 m/s, 2.0–2.3 m/s, and 2.8–3.0 m/s. The first "day" consisted of pooled samples for June 25 and 26, the second day of samples collected July 2, and the third day of samples collected July 7, 1986. The two depths

[1]Present address: Finnish Game and Fisheries Research Institute, Post Office Box 202, SF-00151, Helsinki, Finland.

were 0.5 m and 1.0 m. The total number of samples was 56, all taken at the same station.

The sampling station was in the shallow area of the inlet, where the water depth was 1.3–1.5 m. The distance to the shoreline was approximately 300 m. For each sample, the sampler covered a total distance of 500 m, 250 m each in opposite directions. Of the two parallel samplers, one was set at a depth of 0.50 m and the other at 1.0 m. Data on air temperature, wind direction and speed, and hours of sunshine were obtained from the Meteorological Institute, Helsinki. All meteorological data were obtained for an observation point (Helsinki Kaisaniemi) approximately 3 km from the sampling area, except the hours of sunshine, which refer to an observation point (Vantaa) approximately 15 km from the sampling area.

The samples of larvae were preserved immediately after sampling in 4% formaldehyde solution. In the laboratory, larval European smelt were identified, separated from other larvae, measured to the nearest millimeter, and counted.

The results were analyzed by considering four length-groups of larvae: 6–12, 13–18, 19–24, and 25–37 mm. The first group represented larvae before fin differentiation. The second group consisted of larvae in a stage of intensive fin development. In the third group, the rest of the fins were developed, and in the fourth group, larval features gradually disappeared.

The data were transformed to $\log_e$ (number of larvae in a tow + 1) and analyzed by the General Linear Model Procedure of the SAS statistical package (SAS 1985). By using a log-transformation, we assume that the catching of larvae is a multiplicative process in which the catch is determined by the different factors in a multiplicative fashion. Thus we assumed that the variable $Y = \log_e$ (number of larvae + 1) could be analyzed with a nested generalized linear model in which the factors are nested within the variable length:

$$Y_{ijke} = E(Y_{ijke}) + \epsilon_{ijke}; \quad (1)$$

Y_{ijke} = transformed number of larvae caught on the ith sampling date, at the jth depth with the kth sampler speed within the eth length-group;

$E(Y_{ijke})$ = expected value of Y_{ijke};

ϵ_{ijke} = normal deviation from expectation.

The linear predictor η is

$$\eta_{ijke} = \mu + \text{day}_i\,(\text{length}_e) + \text{depth}_j\,(\text{length}_e) + \text{speed}_k\,(\text{length}_e) + \text{interactions}\,(\text{length}_e); \quad (2)$$

μ = overall mean;

day = effect of sampling date;

depth = effect of depth;

speed = effect of sampler speed;

length = length-group (nesting variable).

Because the deviance was assumed to be normal, the link between the linear predictor and the expected value is the identity link function, and

$$E(Y) = \eta. \quad (3)$$

The standardized residuals between the observed and predicted values were plotted against the predicted values in order to check for trends and deviations from normality.

The frequency of zero catches and catches greater than 10 and greater than 20 individuals in a tow were analyzed separately as functions of day, depth, and speed of the sampler. The association between the variables was measured by the Cochran–Mantel–Haenszel general association statistic (SAS 1985). The null hypothesis was that there was no association between the catch frequencies and depth or sampler speed at any of the strata that were defined by sampling day and depth or speed. Depth was used as a stratifying variable when speed was analyzed in order to avoid confounding of the speed effect by the depth effect. For the same reason, speed strata were included in the analysis of depth effects. All the analyses were performed separately for each length-group.

Results

Variability of Catches

Larvae 6–37 mm long (total length) were caught. During the first day, the densities in each 2-mm length-class remained below 0.5 larvae/m^3, and the mode of the length distribution was in the class of 10–11 mm. During the second and third days, more than 1.0 larvae/m^3 were observed, and the modes of the length distribution shifted to 20–21 mm and 22–23 mm, respectively (Figure 1).

The variance of catches of larvae was high relative to the mean in all length-groups due to the non-normal distributions of the catches (Figure 2). The variance-to-mean ratios were high in the different length-groups and within any sampling day. The variance was a linearly increasing function of the mean for the untransformed data, but was stabilized by the $\log_e$ transformation (Table 1).

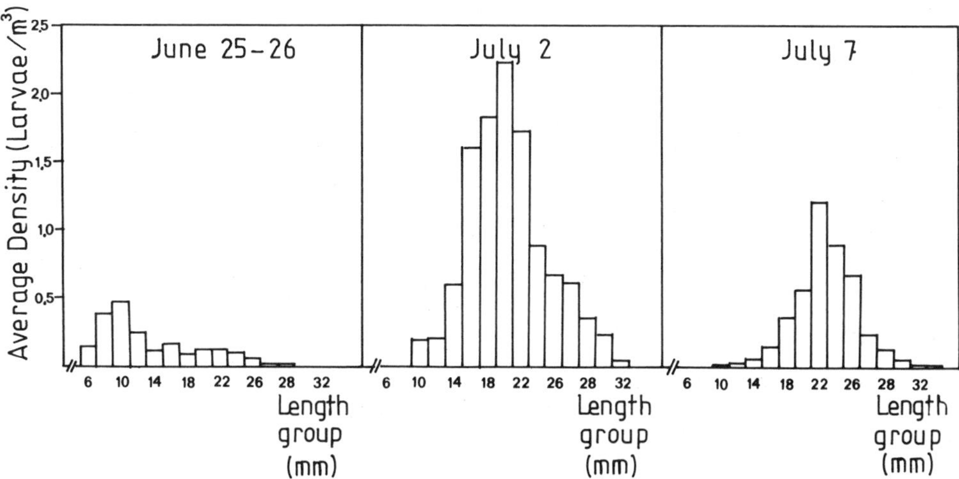

FIGURE 1.—Length distributions of European smelt larvae in 2-mm length-classes during sampling days I (June 25–26), II (July 2), and III (July 7) given as observed average densities (larvae/m^3). The sampler speed was approximately 2 m/s.

Factors Determining the Catches

The day, depth, and speed effects significantly influenced the catches of larvae ($P < 0.01$; Table 2). The speed effect was the smallest, and plots of the transformed catches against speed show that sampler speed had virtually no effect on the catches in the smallest length-group (Figure 3). Of the interaction terms, day × depth proved significant, whereas the other pairwise interactions accounted for an insignificant part of the variance ($P > 0.3$) and were excluded from the model. Overall the three factors and the day × depth interaction explained about 88% of the total variation of the log-transformed counts. A plot of the residuals against fitted values revealed no trends, which suggested that the model adequately described the data. The residuals were also normally distributed for all length-groups combined, but within length-groups some skewness was observable. Thus the number of larvae in the second length-group (13–18 mm) was more often underestimated than overestimated.

Zero Catches and Extreme Catches

In all length-groups except the smallest (6–12 mm), the frequency of zero catches decreased

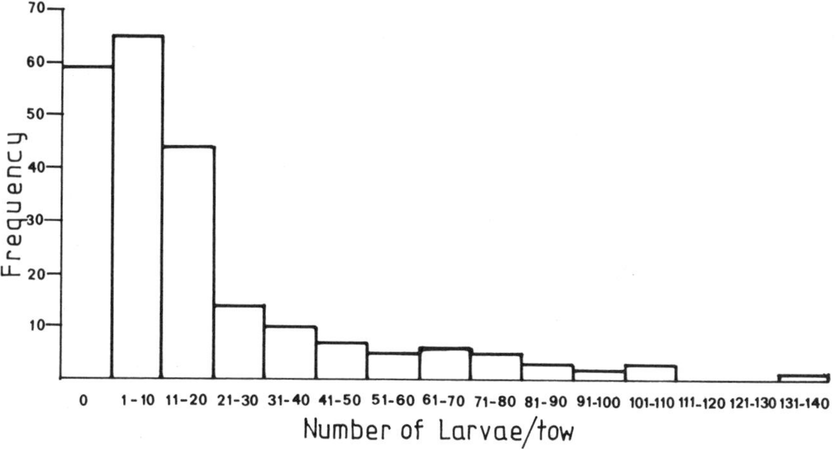

FIGURE 2.—Frequency distribution of larval European smelt catches pooled over sampling days, sampler speeds, depths, and larval length. The skewed distribution suggests that a logarithmic transformation of the data is appropriate.

TABLE 1.—Summary statistics for sampling of larval European smelt in the Vanhankaupunginlahti inlet, northern Baltic Sea, pooled over depth and sampler speed.

Sampling day[a]	Larval length (mm)	Catch (larvae/tow)			$\log_e$ (catch + 1)		
		Mean	Var	CV[b]	Mean	Var	N
I	6–12	11	106	95	1.89	1.64	21
	13–18	3	12	104	1.10	0.85	21
	19–24	2	6	111	0.88	0.69	21
	25–37	1	2	140	0.50	0.41	21
II	6–12	3	14	110	1.07	0.96	16
	13–18	40	1,521	97	2.90	2.47	16
	19–24	53	1,661	77	3.49	1.73	16
	25–37	25	517	90	2.73	1.52	16
III	6–12	<1	<1	318	0.09	0.08	19
	13–18	6	12	61	1.73	0.37	19
	19–24	28	459	75	3.12	0.58	19
	25–37	18	247	87	2.57	0.89	19
I–III	6–12	5	65	158	1.04	1.47	56
	13–18	15	693	179	1.83	1.64	56
	19–24	26	1,035	125	2.39	2.31	56
	25–37	14	329	132	1.84	1.96	56

[a]Day I = June 25–26, 1986; day II = July 2, 1986; day III = July 7, 1986.
[b]Coefficient of variation: CV = 100 · SD/mean; means, variances and CVs have been rounded from decimal fractions.

from June 26 to July 7. Zero catches were recorded in all speed classes, but the proportion of zero catches was highest in the first two speed classes in which the sampler speed was less than 1.8 m/s (Figure 4). The observations of zero frequencies in the shortest length-group were not associated with the sampling speed, but in the other length-groups there appeared to be an association with speed (Table 3).

Most of the zero catches were observed at a depth of 0.5 m. This tendency was clearest during the first two days in the length-groups 1 and 2 (larvae of 6–18 mm), when all zero catches were recorded closer to the surface (Figure 4). The general association statistics also revealed that the connection between zero frequencies and depth became weaker for larger larvae (Table 3).

The catches of more than 20 larvae in a tow represented the extreme alternative to the zero catches. The greatest number of large catches was made during the second day in all length-groups except the smallest. The role of speed in determining extreme catches was minor, as demonstrated by the statistical analysis (Table 3). The sampler speed was, however, significant in determining catches of more than 10 larvae (Table 3; Figure 4).

There was a clear depth segregation during the second day, when most of the large catches were at 1 m, but by the third day this pattern had largely broken down (Figure 4). The same pattern was observed for both the catches of more than 10 and the catches of more than 20 larvae in a tow, and was also revealed by the average transformed

TABLE 2.—Results of fitting a generalized linear model to the $\log_e$(catch + 1)-transformed data for larval European smelt sampled in the Vanhankaupunginlahti inlet, northern Baltic Sea. In the model, larval length was used as a nesting variable.

Source	df	Sum of squares	Mean square	F	P	r^2
Model	35	404.2	11.6	40.8	0.0001	0.88
Day (length)	8	184.1	23.0	81.4	0.0001	
Depth (length)	4	62.0	15.5	54.9	0.0001	
Speed (length)	12	39.7	3.3	11.7	0.0001	
Day × depth (length)	8	54.7	6.8	24.2	0.0001	
Error	188	53.16	0.283			
Total	223	457.3				

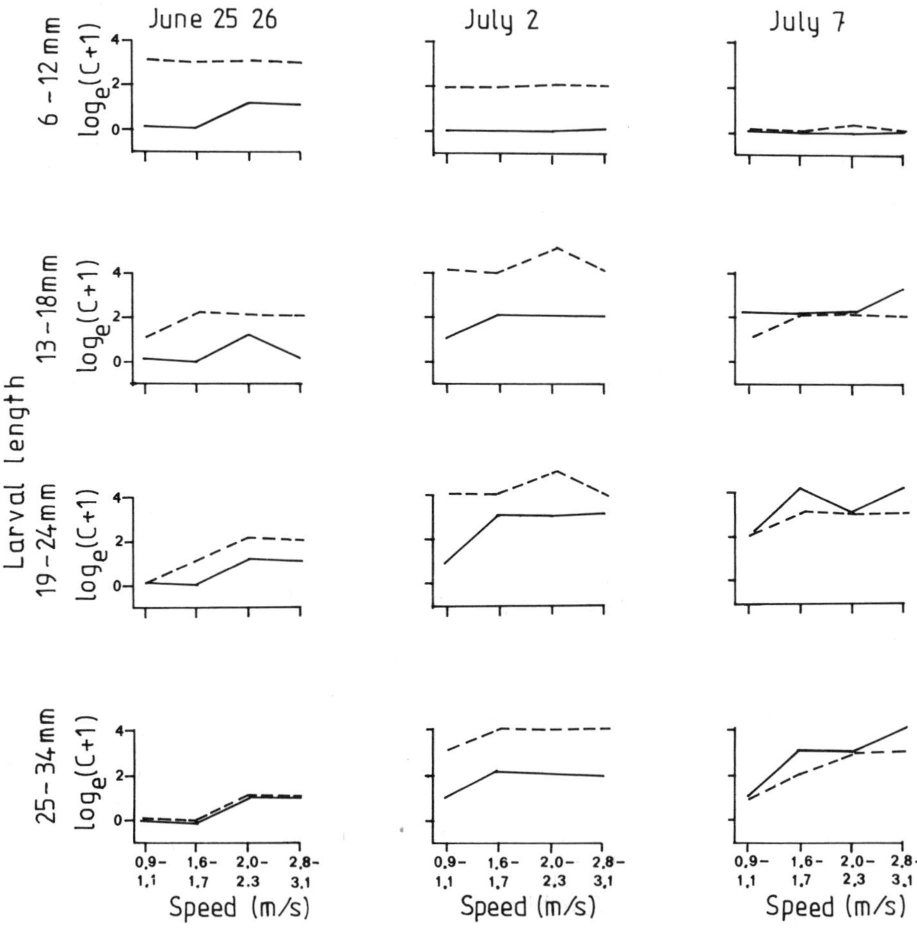

FIGURE 3.—Average $\log_e$-transformed catch of European smelt as a function of sampler speed during sampling days I (June 25–26), II (July 2), and III (July 7) for length-groups 6–34 mm. The solid line marks the 0.5-m depth samples and the broken line the 1-m depth samples.

catches (Figure 3). The association statistics demonstrated that the role of depth in determining large and extreme catches changed rather dramatically when the larvae reached the third length-group (19–24 mm; Table 3).

Discussion

Sampling Variability

The observation that the variability is large in sampling fish and other animals is an old one (e.g., Winsor and Walford 1936; Silliman 1946). In many cases, the negative binomial distribution has adequately reflected the distribution of individuals in catches because of the contagious distribution of the population being sampled (Moyle and Lound 1960; Roessler 1965).

In this experiment, highly varying densities were recorded at the same sampling station. Similar results were obtained by Cole and MacMillan (1984) in sampling larval fish in Lake Erie. This suggests that a significant part of the variability is due to the behavior and movements of the larvae.

The role of sampler speed should be judged against this background. Both the central tendency and the frequency of zero catches depend on the speed of the sampler when the larvae are larger than 12 mm (Figure 3; Table 3). Thus, a wrongly chosen sampler speed may bias a log-linear analysis of zero and nonzero catches, as suggested by Hamley et al. (1983). The Gulf sampler should, therefore, have a minimum speed of approximately 2 m/s in order to avoid biases due to sampling speed. This is slightly less than the

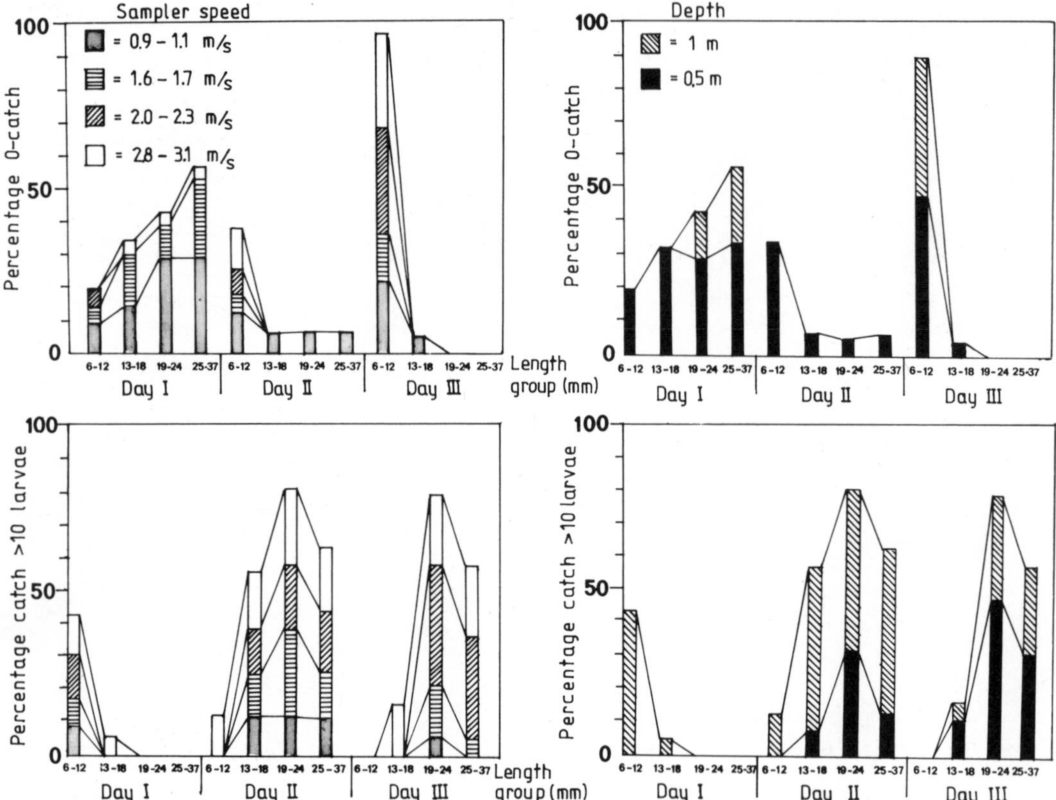

FIGURE 4.—Percentages of tows giving zero and more than 10 European smelt larvae per tow as functions of sampling day, larval length, sampler speed, and depth.

minimum speed of 2.6 m/s recommended by Gehringer (1962) but higher than the speed at which many other nets are towed (e.g., Thayer et al. 1983). Even with this precaution, the Gulf sampler probably underestimates the density of large (>24 mm) larvae. Because the net avoidance capacity of the larvae increases gradually with size, as noted by Noble (1970) and further confirmed by the significant association between sampler speed and extreme catches (Table 3), the density of large larvae will, in general, be underestimated.

Origin of the Variability

One of the main questions raised by this study is why the abundance of fish larvae varies so much even at the same sampling station, as manifested in the very large day effect (Table 2).

The development, growth, and mortality of larvae may cause some of the day-to-day variation. This could be seen in the increasing frequency of zero catches in length-group 1 (6–12 mm) and the decreasing frequency of zero catches in the other length-groups from the first to the last sampling day (Figure 4). Some of the changes in the length distribution (Figure 1) were also due to the growth and mortality of the larvae. More important than simple growth out of one length-group and into the next, however, was the movements of the larvae and ontogenic changes in behavior.

Larval Behavior as a Cause of Variability

The movements of the larvae were affected by currents, which in our case were mainly wind-driven. The high densities during the second and third days relative to the first day (Figure 1) were caused by a movement of larger larvae into the sampling area. The driving force was the wind. The wind speed changed from the first day to the second from 2.8 to 6.8 m/s. The northwesterly wind was apparently strong enough to transport larger larvae from other areas of the bay. From the second to the third day, the wind speed decreased to 3.2 m/s and the larval densities

TABLE 3.—Test for association between sampler speed or sampling depth and frequency of zero, large, and extreme catches of European smelt larvae. The association test is based on the Cochran–Mantel–Haenszel statistic (CMH). The confounding effects of day and speed or depth have been removed by studying the association at the different levels of day and speed or depth. The association test for speed are based on 3 df and the association test for depth on 1 df. The total sample size for all associations was 56.

Larval length (mm)	Association with speed		Association with depth	
	CMH	*P*	CMH	*P*
		Frequency of zero catches		
6–12	3.15	0.37	10.47	0.001
13–18	10.28	0.016	7.42	0.006
19–24	14.90	0.002	4.44	0.035
25–34	14.96	0.002	3.00	0.083
		Frequency of large catches (>10 larvae/tow)		
6–12	6.62	0.085	14.80	<0.001
13–18	19.62	<0.001	7.00	0.008
19–24	13.19	0.004	0.23	0.63
25–34	11.08	0.011	3.91	0.048
		Frequency of extreme catches (>20 larvae/tow)		
6–12	5.11	0.16	5.63	0.018
13–18	Singular covariance matrix		12.00	0.001
19–24	5.58	0.13	1.31	0.25
25–34	8.70	0.034	1.28	0.26

remained similar to those recorded during the second day (Figure 3).

The ontogenic changes in behavior were manifested in the changing relationship to depth. Depth was an important variable for larvae of 6–18 mm, but less so for larger larvae. Thus, the importance of depth in determining catch frequencies decreased with increasing size of the larvae (Table 3). The changing role of depth reflected the increasing swimming ability of the larvae and more extensive foraging in the water column as the larvae developed.

The vertical distribution was not, though, very rigid, and the statistical analysis revealed a significant day × depth interaction (Table 2). It arose because the vertical distribution changed in response to environmental stimuli (Figure 3). Of these stimuli, light is probably one of the most important. During the third sampling day, the sky was overcast. During the first and the second days, 16.8 and 10.8 h of sunshine were recorded, respectively, whereas the third day had only 2.1 h. On the third day, the depth pattern broke down (Figures 3, 4). A similar observation was made by Schnack (1974) for larvae of Atlantic herring *Clupea harengus harengus*.

The role of responses to environmental stimuli may obviously be exaggerated in a shallow area such as the Vanhankaupunginlahti, but similar patterns in behavior could be important in deeper waters as well. Patterns of larval distribution and densities may therefore appear inconsistent, but this inconsistency reflects an inability to take larval behavior into account in the sampling programs rather than truly unpredictable changes in abundance.

In conclusion, one can note that, of the factors affecting the variability of samples of larval European smelt, the most technical one and the one that is easy to adjust—sampler speed—generally explained the smallest part of the total variation. The largest part of the variability was due to the day effect, and this variability cannot be reduced or even correctly interpreted unless we have a profound understanding of larval behavior.

Acknowledgments

Sari Laurila and Antti Koli skillfully handled the sampling. Comments by Eero Aro, Raimo Parmanne, the editors, and four anonymous referees improved the presentation substantially. Pia Åkerman and Ritva Lehtonen drew the figures. The work was financially supported by the Finnish Game and Fisheries Research Institute and the University of Helsinki (Hermann Rosenberg Grant).

References

Bjørke, H., O. Dragesund, and Ø. Ulltang. 1974. Efficiency test on four high-speed plankton samplers. Pages 183–200 *in* J. H. S. Blaxter, editor. The early life history of fish. Springer-Verlag, New York.

Bridger, J. P. 1958. On efficiency tests made with a modified Gulf III high-speed tow-net. Journal du Conseil, Conseil International pour l'Exploration de la Mer 23:357–365.

Cole, R. A., and J. R. MacMillan. 1984. Sampling larval fish in the littoral zone of western Lake Erie. Journal of Great Lakes Research 10:15–27.

Colton, J. B. Jr., J. R. Green, R. R. Byron, and J. L. Frisella. 1980. Bongo net retention rates as affected by towing speed and mesh size. Canadian Journal of Fisheries and Aquatic Sciences 37:606–623.

Gehringer, J. W. 1962. The Gulf III and other modern high-speed plankton samplers. Rapports et Procès-Verbaux des Réunions, Conseil Permanent International pour l'Exploration de la Mer 153:19–22.

Hamley, J. M., T. P. Howley, and A. L. Punhani. 1983. Estimating larval fish abundances from plankton net catches in Long Point Bay, Lake Erie, in 1971–78. Journal of Great Lakes Research 9:452–467.

Hudd, R., M. Hildén, L. Urho, M.-B. Axell, and L.-A. Jåfs. 1984. Fiskeriundersökning av Kyro älvs mynnings- och influensområde 1980–82. [Fishery investigation in 1980–1982 of the Kyrönjoki estuary and its influence area in the Northern Quark of the Baltic Sea.] National Board of Waters, Report 242B:1–277, Helsinki, Finland. (In Swedish with English summary.)

Moyle, J. B., and R. Lound. 1960. Confidence limits associated with means and medians of series of net catches. Transactions of the American Fisheries Society 89:53–58.

Nellen, W., and D. Schnack. 1975. Sampling problems and methods of fish eggs and larvae investigations with special reference to inland water. EIFAC (European Island Fisheries Advisory Commission) Technical Paper 23 (Supplement 1):538–551.

Noble, R. L. 1970. Evaluation of the Miller high-speed sampler for sampling yellow perch and walleye fry. Journal of the Fisheries Research Board of Canada 27:1033–1044.

Roessler, M. 1965. An analysis of the variability of fish populations taken by otter trawl in Biscayne Bay, Florida. Transactions of the American Fisheries Society 94:311–318.

SAS. 1985. SAS/STAT guide for personal computers, version 6. SAS Institute, Cary, North Carolina.

Schnack, D. 1974. On the biology of herring larvae in the Schlei Fjord, western Baltic. Rapports et Procès-Verbaux des Réunions, Conseil Permanent International pour l'Exploration de la Mer 166:114–123.

Silliman, R. P. 1946. A study of variability in plankton tow net catches of Pacific pilchard (*Sardinops caerulea*) eggs. Journal of Marine Research 6:74–83.

Thayer, G. W., D. R. Colby, M. A. Kjelson, and M. P. Weinstein. 1983. Estimates of larval-fish abundance: diurnal variation and influences of sampling gear and towing speed. Transactions of the American Fisheries Society 112:272–279.

Winsor, C. P., and L. A. Walford. 1936. Sampling variations in the use of plankton nets. Journal du Conseil, Conseil International pour l'Exploration de le Mer 11:190–204.